Sustainable strategies for managing *Brassica napus* (oilseed rape) resistance to *Leptosphaeria maculans* (phoma stem canker)

Sustainable strategies for managing *Brassica napus* (oilseed rape) resistance to *Leptosphaeria maculans* (phoma stem canker)

Under the aegis of a European-Australian workshop held at INRA, Versailles, France, in association with the EU SECURE project (QLK5-CT-2002-01813)

Edited by
B.D.L. Fitt, N. Evans, B.J. Howlett and B.M. Cooke

Reprinted from *European Journal of Plant Pathology*, Volume 114 Issue 1, 2006

Springer

A C.I.P catalogue record for this book is available from the library of Congress

ISBN 1-4020-4524-7 (HB)
ISBN 1-4020-4525-5 (e-book)

Published by Springer,
P.O. Box 17, 3300 AA, Dordrecht, The Netherlands

Printed on acid-free paper

Cover photos:
Phoma stem canker of oliseed rape (*Brassica napus*) caused by *Leptosphaeria maculans* and *L. biglobosa*. From top to bottom: Symptoms on leaves ans stems; world-wide distribution; stem cankers of differing severities; pathogen life cycle in Europe in relation to host resistance. Full details in Fitt *et al.* (pp. 3–15)

Printed in the Netherlands

Contents

European Journal of Plant Pathology (2006) 114: 1
DOI 10.1007/s10658-005-8307-6

Foreword

Sustainable strategies for managing *Brassica napus* (oilseed rape) resistance to *Leptosphaeria maculans* (phoma stem canker)

The interaction between the fungus *Leptosphaeria maculans* and oilseed rape *(Brassica napus)* is becoming an excellent model system for studying genetics of host–pathogen interactions. *Leptosphaeria maculans* causes phoma stem canker (blackleg) on oilseed rape and other Brassica crops worldwide. Recently, application of molecular techniques has led to increased understanding of the genetics of this hemibiotrophic interaction. The complete sequences of the genomes of *L. maculans* and *B. rapa* (comprising the *Brassica* A genome) will be available soon. This will provide new opportunities to investigate basic metabolic pathways in the host and the pathogen, and detailed knowledge of the disease process.

Worldwide, the major strategy for control of phoma stem canker is the use of cultivars with resistance to *L. maculans*. However, serious epidemics have occurred recently in Australia and Europe when *L. maculans* populations changed such that major gene resistance in oilseed rape was overcome. Thus there is an urgent need to find and deploy sources of resistance to *L. maculans* in a manner that enhances their durability.

This topic was addressed at a workshop (phoma stem canker durable resistance workshop, 13 September 2004) attended by plant pathologists from Australia and Europe, which was held at INRA (Versailles, France) immediately before the second annual meeting of the European Union-funded SECURE (QLK5-CT-2002-01813) project. This special issue of the European Journal of Plant Pathology is based on papers presented at the workshop, with additional contributions from scientists in Canada, Australia and Europe who were not able to attend. To provide an international perspective on each topic, authors of most of the papers are from several countries. We hope that these papers will provide researchers with a synthesis of the recent studies relating to strategies for management of resistance genes to provide effective control of *L. maculans*, and will stimulate further research on this important model system.

BRUCE D.L. FITT
NEAL EVANS
BARBARA J. HOWLETT
MIKE COOKE

European Journal of Plant Pathology (2006) 114: 3–15
DOI 10.1007/s10658-005-2233-5

World-wide importance of phoma stem canker (*Leptosphaeria maculans* and *L. biglobosa*) on oilseed rape (*Brassica napus*)

B.D.L. Fitt[1,*], H. Brun[2], M.J. Barbetti[3] and S.R. Rimmer[4]
[1]*Rothamsted Research, Harpenden, AL5 2JQ, Herts, UK;* [2]*INRA, UMR Bi03P, Domaine de la Motte, BP 35327, F-35653, Le Rheu Cedex, France;* [3]*School of Plant Biology, University of Western Australia, 35 Stirling Highway, WA 6009, Crawley, Australia;* [4]*Agriculture and Agri-Food Canada, Saskatoon Research Centre, 107 Science Place, S7N OX2, Saskatoon, Canada;* **Author for correspondence (Phone: +44-1582-763133; Fax: +44-1582-760981; E-mail: bruce.fitt@bbsrc.ac.uk)*

Accepted 11 August 2005

Key words: blackleg, durable resistance, host–pathogen genetics, invasive species, resistance genes, species co-existence

Abstract

Phoma stem canker is an internationally important disease of oilseed rape (*Brassica napus*, canola, rapeseed), causing serious losses in Europe, Australia and North America. UK losses of €56M per season are estimated using national disease survey data and a yield loss formula. Phoma stem canker pathogen populations comprise two main species, *Leptosphaeria maculans*, associated with damaging stem base cankers, and *Leptosphaeria biglobosa*, often associated with less damaging upper stem lesions. Both major gene and quantitative trait loci mediated resistance to *L. maculans* have been identified in *B. napus*, but little is known about resistance to *L. biglobosa*. *Leptosphaeria maculans*, which has spread into areas in North America and eastern Europe where only *L. biglobosa* was previously identified, now poses a threat to large areas of oilseed rape production in Asia. Epidemics are initiated by air-borne ascospores; major gene resistance to initial infection by *L. maculans* operates in the leaf lamina of *B. napus*. It is not clear whether the quantitative trait loci involved in the resistance to the pathogen that can be assessed only at the end of the season operate in the leaf petioles or stems. In countries where serious phoma stem canker epidemics occur, a minimum standard for resistance to *L. maculans* is included in national systems for registration of cultivars. This review provides a background to a series of papers on improving strategies for managing *B. napus* resistance to *L. maculans*, which is a model system for studying genetic interactions between hemi-biotrophic pathogens and their hosts.

Introduction

Phoma stem canker (blackleg) is a disease of world-wide importance on oilseed rape (*Brassica napus*, canola, colza, rapeseed, Raps), which can cause serious losses on crops in Europe, Australia and North America (West et al., 2001; Howlett, 2004). The disease is caused by a complex of *Leptosphaeria* species (Mendes-Pereira et al., 2003), the most important of which is *L. maculans*, associated with damaging stem base canker in many countries (West et al., 2001). In Europe and North America, *L. maculans* often co-exists with *L. biglobosa* (West et al., 2002a), which may have evolved from a common ancestor (Gudelj et al., 2004). *Leptosphaeria biglobosa* is associated with upper stem lesions; whilst generally not damaging, they can cause serious losses in countries like Poland with high summer temperatures (Huang et al., 2005).

Basal phoma stem canker (*L. maculans*) can potentially cause total crop loss, for example when highly susceptible Chinese cultivars were grown in Europe (Grezes-Bessett and McCartney, personal communication) or when breakdown of major gene resistance in a susceptible background occurred recently in Australia (Li et al., 2003). This major gene resistance generally operates at the point of entry of *L. maculans* into the plant (cotyledon or leaf), although its effects may last throughout the season because *L. maculans* is a monocyclic pathogen. However, many cultivars grown in countries where *L. maculans* is endemic also have some quantitative background resistance to *L. maculans*, which may operate to impede the progress of the pathogen down the leaf petiole or in the stem tissues, although the genetics is not clearly understood (Rimmer and van den Berg, 1992; Delourme et al., 2004).

Fungicide spray treatments, applied to control stem canker in western Europe in autumn/winter during the leaf spot phase of the disease before the pathogen reaches the stem (West et al., 1999, 2002b), may become impractical if gross margins from growing winter oilseed rape decrease in these countries. Use of fungicide foliar sprays is generally uneconomic outside western Europe, in countries where yields are lower, although fungicides are applied with the seed in Australia and Canada (West et al., 2001). Therefore, for sustainable world-wide production of oilseed rape, strategies need to be developed to manage resistance to *L. maculans* so that it is durable (Rouxel et al., 2003a). This review, which provides the background for a series of papers on developing improved strategies for managing *B. napus* resistance to *L. maculans*, discusses national losses from phoma stem canker, differences between *L. maculans* and *L. biglobosa*, the world-wide spread of *L. maculans*, epidemiology of phoma stem canker in relation to genetics of *B. napus* resistance to *L. maculans* and the role of disease resistance in national systems for registration of oilseed rape cultivars.

National losses from phoma stem canker

Phoma stem canker is now the most serious disease on winter oilseed rape in the UK. Using data from a national (England and Wales) survey, estimates of losses from this disease have increased from *c.* €14M per season in the late 1980s (Fitt et al., 1997) to €56M per season in harvest years 2000–2002 (www.cropmonitor.co.uk) (Figure 1). By contrast, losses from light leaf spot, caused by *Pyrenopeziza brassicae*, the most serious disease of winter oilseed rape in Scotland, have decreased in England and Wales in this period, with estimated losses of *c.* €28M per season in 2000–2002. On a national scale in the UK, both sclerotinia stem rot (*Sclerotinia sclerotiorum*) and dark pod spot (*Alternaria brassicae*) are generally unimportant, with average losses of €2M and €0.4M per season, respectively (Fitt et al., 1997).

These losses are estimated by multiplying England and Wales survey disease incidence data (% plants affected) and appropriate yield loss parameters (Fitt et al., 1997). The survey data are collected by sampling from a stratified series of

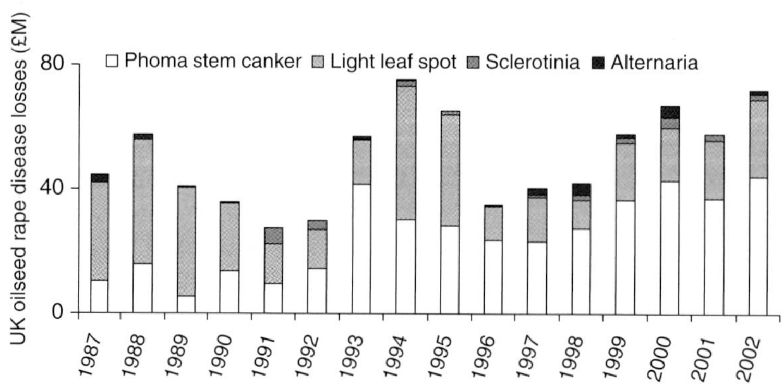

Figure 1. Estimated losses (£1 ≡ €1.4) from diseases (phoma stem canker, light leaf spot, sclerotinia stem rot and alternaria pod spot) in winter oilseed rape in England and Wales, for harvest years 1987–2002, calculated from disease survey data (www.cropmonitor.co.uk) and yield loss coefficients (Fitt et al., 1997).

approximately 100 commercial crops per season (Welham et al., 2004), with numbers of crops sampled proportional to the area of oilseed rape grown in each Defra (UK Department for Environment, Food and Rural Affairs) region. Losses are estimated from data for % plants affected in random samples of 25 plants per crop taken in summer (early July) before harvest. Estimates of yield losses associated with severe epidemics of each of these diseases were based on data from plot experiments with only one disease present, in which fungicides had been used to control this disease. Thus, yield response to fungicide treatment (y) was related to decrease in incidence of phoma stem canker (x) by linear regression ($y = a + bx$). The yield loss coefficient (b) for phoma stem canker was estimated as 0.015 t ha^{-1} for each 1% increase in incidence of the disease. Given the area sown to oilseed rape each season, the national incidence of the disease and the oilseed rape price (estimated as €210 t^{-1}), the yield loss coefficient was used to estimate the loss from phoma stem canker each season to demonstrate trends in the national importance of the disease. Yield loss coefficients relating % yield loss to incidence of phoma stem canker have also been estimated (Zhou et al., 1999); given data for the national average yield in t ha^{-1}, these could also be used to estimate national average yield losses.

Despite the deployment of resistant cultivars, the oilseed rape industry in Australia continues to suffer serious losses from phoma stem canker, as illustrated by losses of €11.3M and €30.1M for the 1998 and 1999 seasons, respectively (Khangura and Barbetti, 2001). In France, losses from phoma stem canker vary between regions and seasons, but generally account for 5 (€36.8M) to 20% (€147M) of the national oilseed rape production (Allard et al., 2002).

Pre-harvest assessments of phoma stem canker can be used, retrospectively, to estimate yield losses from the disease in Europe because most losses are associated with premature death of plants through occlusion of vascular tissues by stem base cankers (West et al., 2001). Although phoma leaf spotting epidemics may be widespread in autumn and winter, such epidemics rarely cause extensive death of plants. If occasional plants are lost, surrounding plants can compensate so that yield is unaffected. By contrast, in Australia, widespread death of seedlings and complete destruction of crops by the disease at any stage from seedling to maturity can occur (Khangura and Barbetti, 2001). In such circumstances, national losses cannot be estimated solely from end-of-season disease surveys and total production of seed.

Differences between *Leptosphaeria maculans* and *L. biglobosa*

Historically, the *L. maculans*/*L. biglobosa* species complex was divided into two groups of isolates, named highly virulent/aggressive and weakly virulent/non-aggressive, from their pathogenicity to oilseed rape stems (Williams and Fitt, 1999). The presence of a non-host specific phytotoxin, sirodesmin PL, in culture filtrates made it possible to divide isolates into Tox$^+$ (producing sirodesmin PL, highly virulent) and Tox^0 (not producing sirodesmin PL, weakly virulent). Moreover, two different RLFP patterns associated with differences in pathogenicity and pigment production in liquid medium lead to classification of isolates into A (highly virulent, Tox$^+$) or B (weakly virulent, Tox^0) groups. B-group isolates are a more complex group than A-group isolates. Indeed, B-group isolates were divided into three subgroups; NA1 (NA, non-aggressive), NA2 and NA3 (Koch et al., 1991).

Under *in vitro* conditions, reproducible differences in pseudothecial morphology, the inability to cross A with B-group single ascospore isolates and crossing of opposite mating types of A with A or B with B suggested that the two groups are different species, named *L. maculans* for A-group isolates and *L. biglobosa* for NA1 B-group isolates (Somda et al., 1997; Shoemaker and Brun, 2001). The two species also differ in germination, growth, pigment diffusion, biochemical traits, molecular patterns and pathogenicity. A study, based on the sequence of the internal transcribed spacer region of the ribosomal DNA repeat, established the relationships between seven members of the species complex. These included *L. maculans* 'brassicae' (A-group), *L. biglobosa* 'brassicae' (NA1 B-group, predominant in Europe) and *L. biglobosa* 'canadiensis' (NA2 B-group, predominant in Canada) (Mendes-Pereira et al., 2003).

Whilst typical symptoms caused by *L. maculans* (phoma leaf spot lesions and stem base cankers) are easily identified, it is more difficult to recognise specific symptoms for *L. biglobosa*. However, *L. biglobosa* leaf lesions generally differ from those of *L. maculans* (Brun et al., 1997; Toscano-Underwood et al., 2001) (Figure 2). Both species are able to survive on stem debris and produce ascospores on unburied debris, but *L. biglobosa* survives longer on unburied debris than on buried debris (Huang et al., 2003a). Under the same conditions, ascospores of *L. maculans* survive longer than those of *L. biglobosa*. Rates of pseudothecial maturation of the two species are similar at 15–20 °C but *L. biglobosa* matures more slowly than *L. maculans* at <10 °C (Toscano-Underwood et al., 2003). In Europe, no yield loss is associated with leaf lesions of either species. On stems, *L. biglobosa* is mainly confined to upper stems (West et al., 2002a), even though both species occur on different stem tissues, including the pith. In France, premature senescence of oilseed rape crops in the absence of phoma stem canker (Brun and Jacques, 1991), associated with a complex of *L. maculans*, *Verticillium longisporum* and *Fusarium* spp., has caused serious yield losses. It is difficult to attribute the losses to specific components of this pathogen complex. More research is also needed to understand effects of *L. biglobosa* on yield and establish relative yield losses caused by *L. maculans* or *L. biglobosa*.

Whilst resistance to *L. maculans*, which may be either major gene or polygenic, has been described (Pilet et al., 2001; Delourme et al., 2004), little is known about resistance to *L. biglobosa*. Nevertheless, some results indicate that genes for resistance to *L. maculans* are not effective against *L. biglobosa*. For example, the genes *Rlm1* in cv. Vivol (Brun et al., 1997) and *Rlm6* in MX lines (not yet commercialised in Europe) both confer resistance to *L. maculans* but not to *L. biglobosa* (Somda et al., 1998; Brun, unpublished results). More research is needed to investigate potential differences in resistance to *L. maculans* in *B. napus* and other crucifers and find sources of resistance to *L. biglobosa*.

World-wide spread of *L. maculans*

L. maculans and *L. biglobosa* have a world-wide distribution, probably due to their transmission in seed of *B. oleracea*, *B. napus*, *B. rapa* and other brassica crops (West et al., 2001). One or other of them is known to occur in Europe (25 countries), Africa (eight countries), Asia (16 countries), North America (Canada, USA, Mexico), central America (five countries), South America (Argentina and Brazil) and Oceania (five countries) (Anon., 2004)

Figure 2. Symptoms of disease on leaves (phoma leaf spot) caused by *L. maculans* (large pale lesions with pycnidia) or *L. biglobosa* (darker lesions, generally smaller) and stems (basal phoma stem canker, *L. maculans* predominant species present; upper stem lesions, *L. biglobosa* predominant species present) of European winter oilseed rape, and cultures of *L. maculans* (no pigment) or *L. biglobosa* (pigment) on potato dextrose agar.

(Figure 3). In most cases, reports do not distinguish between *L. maculans* and *L. biglobosa* or provide information on the brassica crop on which the pathogen was identified. Reports that distinguish between *L. maculans* and *L. biglobosa* are almost entirely based on characteristics of isolates cultured from oilseed rape (*B. napus*).

Leptosphaeria biglobosa 'canadiensis' has been widespread on oilseed rape in Canada since it was first isolated in 1957. *Leptosphaeria maculans* was first isolated from oilseed rape in Saskatchewan in 1975, and subsequently spread to Alberta by 1983 and Manitoba by 1984 (Gugel and Petrie, 1992). Currently, almost all Canadian oilseed rape production is with resistant cultivars. In a survey from 1998 to 2000 in Alberta, Saskatchewan and Manitoba (Keri, Kutcher and Rimmer, unpublished), *L. biglobosa* accounted for 18–48% of the isolates, depending on the year. Both species are widely distributed in the USA (Anon., 2004). *Leptosphaeria maculans* and *L. biglobosa* have recently been reported from Mexico on *B. oleracea* (Moreno-Rico et al., 2001) and Brazil (Fernando and Parks, 2003) and Argentina (Gaetan, 2005) on oilseed rape.

Both *L. maculans* and *L. biglobosa* 'brassicae' occur in France, the UK and Germany, although the relative frequency of the two species differs between locations (West et al., 2001). Until the mid-1990s, phoma stem canker in Poland was almost exclusively associated with *L. biglobosa* (Jedryczka et al., 1994). By 2002, *L. maculans* was widespread on oilseed rape in western Poland, whereas only *L. biglobosa* was found in eastern Poland (Karolewski et al., 2002). Changes in relative frequencies of the two species were also observed in the Czech Republic and Hungary (Szlávik et al., 2003). Thus, there is evidence of an eastward spread of *L. maculans* from western Europe. *Leptosphaeria biglobosa* is established in Russia but *L. maculans* is not (Jedryczka et al., 2002).

Piening et al. (1975) reported severe phoma stem canker on oilseed rape in Kenya from 1972 to 1974 and indicate that the pathogen was present on vegetable brassicas in 1951. From their description of symptoms (severe basal stem cankers), they were probably caused by *L. maculans* and not *L. biglobosa*. *L. maculans*, reported in Natal, South Africa on cabbage crops (Laing, 1986), has probably spread to oilseed rape, introduced into South Africa in 1994 (http://www.arc.agric.za/institutes/ppri/main/news/number60/moth.htm).

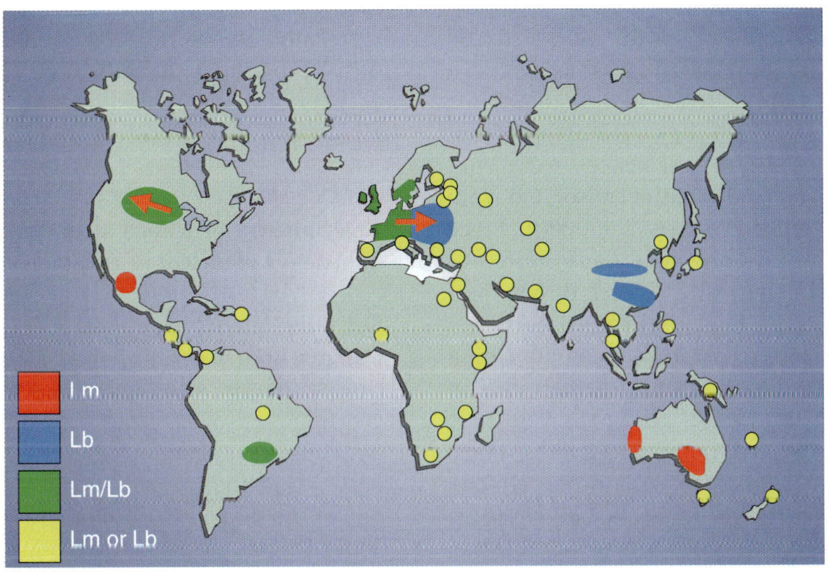

Figure 3. World-wide distribution of *L. maculans* (Lm) and *L. biglobosa* (Lb), showing the direction of spread of *L. maculans* in Canada, where *L. biglobosa* 'canadiensis' was predominant, and eastern Europe (solid arrows), where *L. biglobosa* 'brassicae' was predominant. Areas where populations have been characterised as predominantly *L. maculans* (red), *L. biglobosa* (blue) or a mixture of the two species (green) are indicated by patches. Areas where there have been reports of the pathogens (sometimes only a single report) but the species has not been identified are shown by yellow dots. Based on information in Crop Protection Compendium (Anon., 2004) and other sources available to the authors.

Table 1. Stages in the epidemiology of phoma stem canker (*Leptosphaeria maculans*) in different parts of the world where severe epidemics occur[a]

	Australia	Canada	Europe
Period of ascospore release	Late April–end August	West: May–Aug; Ontario: Sept–Nov, May–Aug	West: Sept–April; East: Sept–Nov, April
Seedling blight (blackleg)	Sporadic outbreaks can severely affect crops (mainly in the west) (June/July)	Occasionally	Uncommon
Phoma leaf lesions	Leaf spots throughout the growing season	West: leaf spots on young or older plants (June, July); Ontario: leaf spots on young winter oilseed rape (Oct–Dec)	West: distinctive leaf spots on young plants, Oct–April; East: little leaf spotting
Phoma stem canker			
Crown canker (stem base)	Most severe phase of disease; can occur at any growth stage	Develops in pre-harvest period (August)	West: most severe phase of disease (May–July); East: rare?
Phoma stem lesions (upper stem)	Observed on stems during and after flowering (Sept–Nov)	Develop in pre-harvest period	Generally more severe in east than west Europe (June/July)
Survival on residues	West: 3–4 years; South–east: 1–3 years	3 years	<2 years

[a]Adapted from West et al. (2001).

Although both *L. maculans* and *L. biglobosa* have been isolated from oilseed rape in Australia (Plummer et al., 1994), the population is almost entirely *L. maculans*. Barrins et al. (2004) found small differences in genetic diversity among isolates according to the cultivar, age of the plants and the region from which they were obtained but populations differing in virulence were not observed. Only *L. biglobosa* has been isolated from oilseed rape in China (West et al., 2000). Since many Chinese cultivars are highly susceptible to *L. maculans* (McCartney and Grezes-Besset, personal communication), this raises the concern that if *L. maculans* isolates are introduced to China considerable damage could result. Furthermore, in China, there are large areas grown to vegetable brassicas. There is a need to improve the resistance to *L. maculans* in Chinese oilseed rape cultivars (*B. napus*) and vegetable brassicas (*B. oleracea, B. rapa*). In the meantime, strict quarantine measures should be employed to ensure that *L. maculans* does not enter China in the next few years. However, two factors relating to crop production practices in China may mitigate the spread and significance of *L. maculans* there. Removal of oilseed rape stem debris from the field after harvest for use as cooking fuel in rural China destroys inoculum. Rotation of oilseed rape with rice involves flooding fields after the oilseed rape harvest, submerging infected residues for long periods. Flooding oilseed rape residues greatly decreased ascospore production after 6 days and almost eliminated it after 10 days (Petrie, 1995).

Epidemiology of phoma stem canker in relation to genetics of *B. napus* resistance to *L. maculans*

Since the oilseed rape growing regions of Europe, Canada and Australia where phoma stem canker causes major economic losses have different growing seasons, types of cultivar resistance, agricultural practices and climates, it is not unexpected that there are differences in the epidemiology of the disease between these areas (West et al., 2001) (Table 1). Wherever phoma stem canker occurs, the air-borne *L. maculans* ascospores are the main source of inoculum (Gladders and Musa, 1980; Salisbury et al., 1995; West et al., 2001) (Figure 4). However, seasonal patterns of ascospore discharge differ between locations and seasons (Khangura and Barbetti, 2001; West et al.,

2002b). Differences in timing of pseudothecial maturity are the main cause of differences in the timing of the start of ascospore discharge (West et al., 1999). Despite this, the main periods of ascospore release in the different countries are predominantly during the late autumn/winter (Gladders and Musa, 1980; West et al., 1999; Salam et al., 2003). In some regions (e.g. Western Australia), ascospore showers often coincide with seedling development (Wherrett et al., 2004). In Western Australia, modelling demonstrated that the dates of both seedling emergence and ascospore development/release are determined by rainfall (Salam et al., 2003).

Maximum yield loss results from ascospore infections that occur at the early seedling stage, when plants are most vulnerable (Barbetti and Khangura, 1999). The role of conidia in the epidemiology of the disease is generally minor in Europe but more important in Western Australia (West et al., 2001). In Australia, there is a good correlation between incidence of cotyledon lesions and subsequent incidence of stem base canker (Li et al., 2005). By contrast, in North America and Europe, cotyledon infection is generally less important (West et al., 2001). In winter oilseed rape, the most damaging stem base cankers are generally associated with phoma leaf spots that developed on leaves three to ten before the onset of rapid stem extension.

Major gene specific resistance to *L. maculans* (Rimmer and van den Berg, 1992; Balesdent et al., 2002) operates when the ascospores infect cotyledons or leaves of seedlings and prevents subsequent spread to the stem and development of cankers (Figure 4). Major gene resistance can be effective for several years under field conditions, provided the corresponding avirulent strains of the pathogen remain prevalent (Rouxel et al., 2003b). However, major gene resistance has broken down in France (Brun et al., 2000; Rouxel et al., 2003b) and Australia (Li et al., 2003; Sprague et al., 2006). Such resistance breakdown is associated with major changes in populations of *L. maculans*. For example, in France, *L. maculans* population changes from avirulence (*AvrLm1*) to virulence (*avrLm1*) to the single dominant *B. napus* resistance gene *Rlm1* between 1990 and 2000 were associated with

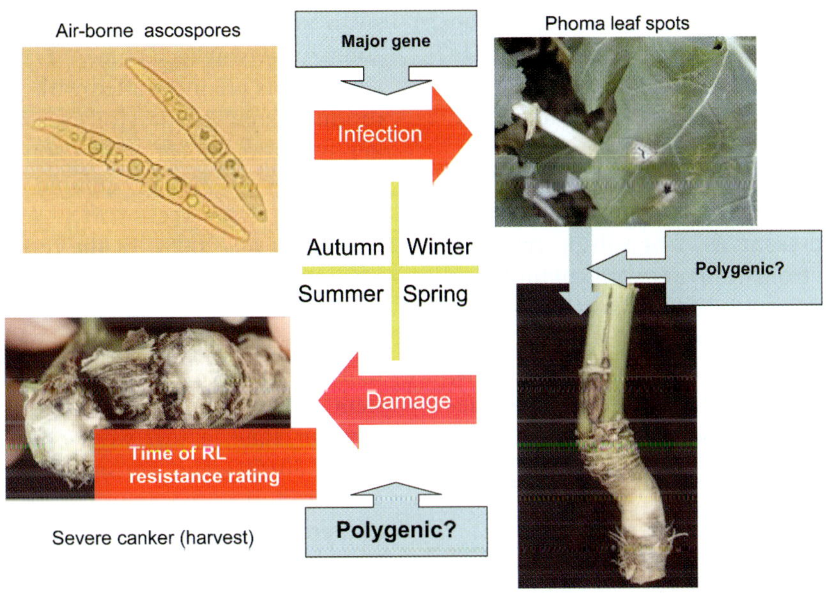

Figure 4. Seasonal cycle of phoma stem canker epidemics in Europe in relation to potential components of oilseed rape (*B. napus*) resistance to *L. maculans*. Epidemics of this monocyclic disease are initiated in autumn (September/October) by air-borne ascospores; the pathogen spreads down the leaf petioles to reach the stem, where stem base cankers or upper stem lesions develop by harvest. A gene-for-gene specific host–pathogen interaction operates at the leaf infection stage but the basis for the background adult plant (quantitative) resistance which operates in the leaf petiole and stem is not known. The UK recommended list (RL) rating for resistance to phoma stem canker (www.hgca.com) is based on assessments of the cross-section of the stem damaged by the pathogen in summer (plants sampled in late June) before harvest (July).

breakdown of this host resistance in commercial crops (Rouxel et al., 2003b). There is also good evidence that the resistance genes *Rlm9*, *Rlm2* and *Rlm4* were rapidly broken down in France after the widespread use of cultivars carrying them (Rouxel et al., 2003b). These studies suggest that a single major gene for resistance operating alone at the leaf infection stage of epidemics is unlikely to be durable.

Cultivars with quantitative resistance, which may operate when the pathogen is spreading down the leaf petiole or into the stem tissues (Figure 4) (West et al., 2001), can be effective in controlling *L. maculans* (Salisbury et al., 1995; Pilet et al., 2001). Use of quantitative resistance in breeding programmes has ensured new cultivars have good background resistance (Delourme et al., 2006). However, quantitative resistance is generally influenced by environmental conditions and its performance can be variable. Quantitatively inherited resistance is likely to be more stable (Pilet et al., 2001) than single gene seedling leaf resistance. Despite this, it is of concern that in Western Australia strains of *L. maculans* may overcome quantitative resistance under glasshouse conditions (Li et al., 2005). As quantitative resistance is controlled by many genetic factors, molecular markers for mapping and characterising quantitative trait loci (QTL) can be used to identify these different genetic backgrounds (Pilet et al., 2001; Delourme et al., 2004).

In Australia, fewer pseudothecia and ascospores were produced on residues from a cultivar with specific resistance from *B. rapa* subsp. *sylvestris* than on residues from cultivars with quantitative resistance (Marcroft et al., 2004). Most ascospores were produced on European winter oilseed rape cultivars with quantitative resistance. Thus the type of resistance deployed may affect reproduction of *L. maculans* and selection for increased virulence.

Importance of resistance to *L. maculans* in national systems for registration of oilseed rape cultivars

In countries where phoma stem canker causes serious epidemics on oilseed rape, there is generally a standard for resistance to *L. maculans* included in the national system for registration of oilseed rape cultivars. For example, in the UK 'recommended list' system for registration of winter oilseed rape cultivars (www.hgca.com) there is an assessment of 'field resistance' to the pathogen (i.e. resistance assessed on adult plants at the end of the season, rather than on cotyledons or leaves of seedlings). Each year the published table of recommended list winter oilseed rape includes 'resistance to stem canker', along with relative gross output, oil content, glucosinolate content and agronomic qualities, such as resistance to lodging and 'resistance to light leaf spot'. The resistance to stem canker or light leaf spot is on a 1 (susceptible) to 9 (resistant) scale, with the minimum standard for resistance to either disease a score of 3. These minimum standards for disease resistance are used in the decision-making progress, alongside other agronomic standards such as minimum lodging resistance, and marketing standards such as low glucosinolate content. Earlier in the selection process, a merit rating for each candidate cultivar is calculated, based on gross output, lodging resistance and resistance to stem canker and light leaf spot.

To assess field resistance to *L. maculans* in the UK, each season a series of 'recommended list' field trials including the candidate cultivars are sown at a range of sites in different parts of the country (www.hgca.com). These include trials where plots are inoculated with winter oilseed rape residues from the previous season with stem canker symptoms, to provide a source of *L. maculans* ascospores to initiate phoma stem canker epidemics a few weeks after sowing in autumn (Huang et al., 2005). In the recommended list trials, contractors assess phoma leaf spot in autumn to confirm that the inoculation has been successful. However, the score (1–9) for field resistance to *L. maculans* is based on assessments of phoma stem canker severity on plants sampled from plots in June, a few weeks before harvest. Currently, 30–50 stems are sampled from each plot in June and the severity of external phoma stem canker at the stem base is assessed (Kenyon et al., 2004) (Figure 5a). Such external assessments may not accurately measure the internal damage to the stem and stems are also cut transversely at the base to record the extent of internal stem blackening (Figure 5b). These external and internal assessments are used to produce an index for stem canker severity, which is inversely related to the resistance rating of the cultivar.

Figure 5. External (a) and increasingly severe internal (b, by comparison with a healthy stem) symptoms of phoma stem canker on winter oilseed rape in the UK, assessed in summer before harvest.

Australia also has a national 'Blackleg Resistance Rating' testing scheme, under which oilseed rape germplasm is screened for resistance to *L. maculans* in a series of phoma stem canker screening nurseries across the oilseed rape growing regions. The main criteria for assessing resistance/susceptibility are based upon the percentage of plants that survive in the presence of high concentrations of *L. maculans* ascospore inoculum. These results are made available to all oilseed rape breeding programmes, pathologists, agronomists and growers throughout Australia on the web site of the Canola Association of Australia. This system assumes that most, if not all, plant loss is from phoma stem canker. In Western Australia, additional assessments are also made of the severity of crown (stem base) canker when flowering has finished and a score is assigned to each cultivar. These additional ratings give an assessment of the impact of phoma stem canker on surviving plants and can readily be related to yield loss. Thus Western Australian growers are provided with a single rating that combines both assessments (Khangura et al., 2003).

In Canada, the 'Western Canadian Canola/Rapeseed Recommending Committee' evaluates candidate cultivars for potential commercial production and forwards recommendations based on seed quality, agronomic performance and phoma stem canker resistance data to the Canadian Food Inspection Agency, which is responsible for cultivar registration. For phoma stem canker resistance, cultivars are evaluated in disease nurseries at a number of sites across western Canada. Cultivar resistance is compared to that of three control cultivars and the overall disease severity at each site is assessed on 'Westar', included as a susceptible control. A minimum of six site-years of data over two years of testing is used to assign each cultivar a rating of susceptible (S), moderately susceptible (MS), moderately resistant (MR) or resistant (R). Currently, cultivars are recommended only if they are MR or R.

In France, the candidate cultivars for registration in the French national list are assessed over a growing season for their resistance to stem base canker in the CTPS (Comité Technique Permanent de la Sélection) network, comprising nine trials in different regions of France (http://www.geves.fr). In each trial, there are eight control cultivars, chosen because they are susceptible/resistant and some of them carry specific *Rlm* genes (*Rlm1*, *Rlm4* and *Rlm7*) present in widely grown commercial cultivars. Each trial is inoculated before plants reach the two-leaf stage with residues infected with *L. maculans*. Leaf lesions are scored in the autumn to confirm that the inoculation has been successful. Forty stem bases are sampled per cultivar and per replicate (four replicates per cultivar per trial) from each plot in June a few weeks before harvest, and the severity of internal phoma stem canker at the stem base is assessed on a scale from 1 (healthy) to 6 (severe disease). The index of candidate cultivars is compared to that of control cultivars. Cultivars that have a greater stem canker index than two susceptible control cultivars are rejected and those that have a stem canker index not significantly different from that of the two resistant control cultivars receive a score up to 1.5, which is included in the cultivar merit rating used in selection of cultivars, alongside yield, seed quality and other agronomic characters.

Discussion

This review provides evidence that *L. maculans*, cause of phoma stem canker of oilseed rape and other brassicas, should be classified as a global invasive species. Not only does it cause a destructive disease under a wide range of climates in Europe, North America and Australia (Salisbury et al., 1995; West et al., 2001), but also it has spread across Canada (Gugel and Petrie, 1992), into Mexico (Moreno-Rico et al., 2001) and is spreading across Poland (Karolewski et al., 2002), into areas where previously only the less damaging *L. biglobosa* was present. *L. maculans* now poses a threat to production of oilseed rape in Asia, especially China where *c.* 8M ha of oilseed rape (*B. napus*) and *c.* 1.7M ha of brassica vegetables are grown annually, mostly by subsistence farmers, and sustainable brassica production depends on use of sources of biodiversity in local wild brassica species. To prevent spread of *L. maculans* into China, a short-term strategy is to hold workshops in China and produce manuals on use of PCR diagnostics to identify the pathogen in imported seed (for plant quarantine staff) and recognise symptoms of the disease in the field (for extension staff), with related publicity. In the longer term, there is a need to introduce durable resistance to *L. maculans* into Chinese cultivars and to understand how interactions between *L. maculans* and *L. biglobosa* (West et al., 2002a) might be exploited to control *L. maculans*. Given the increasing importance of oilseed rape as a crop, to meet the shortfall in world-wide demand for food oil and bio-diesel and the increasing severity of phoma stem canker disease, resistance to *L. maculans* is now a major target in many breeding programmes (Marcroft et al., 2002; Delourme et al., 2004).

However, current resistance breeding programmes are hampered by problems in accurately measuring field resistance to *L. maculans*, which is generally assessed at the end of the season (Figures 4 and 5). These methods for assessing phoma stem canker are time-consuming, expensive and technically difficult, because it is not always easy to distinguish symptoms of phoma stem canker from those of other diseases (Kenyon et al., 2004). Although they provide good assessments of the phase of the disease which affects yield, there is scope for improving them, using new understanding of the epidemiology of the disease (West et al., 1999, 2001; Huang et al., 2003b, 2005) and the genetics of the *L. maculans*/*B. napus* interaction (Balesdent et al., 2002; Delourme et al., 2004) and new PCR based methods for quantifying *L. maculans* DNA in infected tissues (Kenyon et al., 2004). End-of-season assessments provide no information about the components of 'field resistance' that have been operating during the period since the initiation of epidemics by air-borne ascospores impacting on leaves of the seedling crop (West et al., 2001; Huang et al., 2005). For example, spray timing experiments have provided indirect evidence that the rate of progress of *L. maculans* down the leaf petiole may differ between cultivars with different scores for resistance to *L. maculans* (Thomas and Wedgwood, 1998). Furthermore, a good relationship has been observed between amounts of *L. maculans* DNA in leaf petiole samples in November (assessed by quantitative PCR) and resistance scores of cultivars at the end of the season in June (Kenyon et al., 2004). This suggests that quantitative PCR may provide an assessment of resistance that is more reliable, less time-consuming and several months earlier than the pre-harvest assessment methods currently used.

Recent problems with breakdown of major gene resistance to *L. maculans* (Brun et al., 2000), associated with severe phoma stem canker epidemics in Australia (Li et al., 2003) have emphasised the need to develop strategies for deployment of durable resistance, through resistance breeding and disease management programmes (Sprague et al., 2006). Durability of resistance depends on factors such as the type of resistance and its genetic background, type of pathogen and its plasticity, area of crop grown and climate. Durable resistance is difficult to produce against fungal pathogens, such as *L. maculans*, where widespread air-borne ascospore dispersal and sexual recombination occurs in crops (McDonald and Linde, 2002). For breeding programmes, it is probably necessary to focus on field resistance, associated with QTL, which may be difficult to select for because it is under polygenic control (Pilet et al., 2001; Delourme et al., 2004). Major genes for resistance to *L. maculans* operating when the pathogen attempts to infect the leaf may be more durable if they are set in a field resistance background than if they are set in a susceptible background, as they were when resistance breakdown occurred recently in Australia (Li et al., 2003). The durability of major gene resistance may be increased by diversification schemes, which classify the current commercial cultivars by the resistance genes they carry, to guide strategies for deployment of these genes (Gladders et al., 2006). Modelling the effects of different deployment strategies in both space (pattern of areas sown to cultivars with different genes) and time (seasonal pattern of deployment), in relation to different measures of durability of resistance (van den Bosch and Gilligan, 2003), can be used to guide advice on effectiveness of different proposed deployment strategies (Pietravalle et al., 2006).

Many of the factors which relate to strategies for managing *B. napus* (oilseed rape) resistance to *L. maculans* (phoma stem canker) can also be applied to other host-pathogen systems. The availability of an extensive range of host and pathogen tools and resources make *L. maculans*–*B. napus* an excellent model system for studying the genetics of host-pathogen interactions for hemi-biotrophic pathogens. The sequencing of the *L. maculans* genome (Rouxel and Balesdent, 2005; Kuhn et al., 2006) and the subsequent availability of minisatellite markers (Eckert et al., 2005a; Rouxel and Balesdent, 2005) will greatly facilitate work on the genetics of pathogen populations. Furthermore, the labelling of *L. maculans* with both GFP (Sexton and Howlett, 2001) and DsRed (Eckert et al., 2005b) reporter genes and the development of real-time PCR (Kenyon et al., 2004) will enable the important symptomless phase in development of epidemics (spread from the leaf down the petiole to the stem) to be studied both qualitatively and quantitatively. The production at INRA Versailles of ascospores of near-isogenic isolates of *L. maculans* differing at specific avirulence loci (*AvrLm1*, *avrLm1*; *AvrLm4*, *avrLm4*; *AvrLm6*, *avrLm6*) provides an excellent opportunity to study fitness deficits associated with loss of avirulence (Huang et al., 2006) as a predictor of durability of resistance genes (Leach et al., 2001). Furthermore, the development at INRA Rennes of near-isogenic lines of *B. napus* with/without *Rlm6* (corresponding to *AvrLm6*), set in a susceptible background (Eurol, Eurol MX) or an adult plant resistant background (Darmor, Darmor MX) (Delourme et al., 2006) provides a unique opportunity for using the *L. maculans*/*B. napus* system to assess whether the durability of major gene resistance is increased by incorporating it into cultivars with good quantitative resistance. These subjects are covered by papers in this special issue of the *European Journal of Plant Pathology*.

Acknowledgements

We are grateful for funding from the UK Biotechnology and Biological Sciences Research Council and Department for Environment, Food and Rural Affairs (ARO 703, OREGIN), the European Union (QLK5-CT-2002-01813, SECURE), the Australian Grains Research and Development Corporation, the French Institut National de la Recherche Agronomique and Agriculture and Agri-Food Canada. We thank Neal Evans for help with the Figures and Peter Gladders for commenting on the text of this paper.

References

Allard LM, Brun H, Jouffret P, Lagarde F, Penaud A, Pinochet X, Simonin P and Taverne M (2002) Les maladies du colza. Points Techniques du CETIOM, 80 pp

Anon. (2004) Crop Protection Compendium Database, CAB International, Wallingford, UK

Balesdent MH, Attard A, Kuhn AL and Rouxel T (2002) New avirulence genes in the phytopathogenic fungus *Leptosphaeria maculans*. Phytopathology 92: 1122–1133

Barbetti, MJ and Khangura RK (1999) Managing blackleg in the disease-prone environment of Western Australia. Proceedings of the 10th International Rapeseed Congress, 1999, Canberra, Australia. http://www.regional.org.au/papers/index.htm.

Barrins JM, Ades PK, Salisbury PA and Howlett BJ (2004) Genetic diversity of Australian isolates of *Leptosphaeria maculans*, the fungus that causes blackleg of canola (*Brassica napus*). Australasian Plant Pathology 33: 529–536

Brun H and Jacques MA (1991) Le dessèchement prématuré des pieds de colza avant la récolte. Quelques symptômes et agents pathogènes associés. La Défense des Végétaux 262: 7–12

Brun H, Levivier S, Eber F, Renard M and Chevre AM (1997) Electrophoretic analysis of natural populations of *Leptosphaeria maculans* directly from leaf lesions. Plant Pathology 46: 147–154

Brun H, Levivier S, Somda I, Ruer D, Renard M and Chèvre AM (2000) A field method for evaluating the potential durability of new resistance sources: application to the *Leptosphaeria maculans*–*Brassica napus* pathogen systems. Phytopathology 90: 961–966

Delourme R, Pilet-Nayel ML, Archipiano M, Horvais R, Tanguay X, Rouxel T, Brun H, Renard M and Balesdent MH (2004) A cluster of major specific resistance genes to *Leptosphaeria maculans* in *Brassica napus*. Phytopathology 94: 578–583

Delourme R, Chevre AM, Brun H, Rouxel T, Balesdent MH, Dias JS, Salisbury P, Renard M and Rimmer SR (2006) Major genes and polygenic resistance to *Leptosphaeria maculans* in *Brassica napus*. European Journal of Plant Pathology 114: 41–52

Eckert M, Gout L, Rouxel T, Blaise F, Jedryczka M, Fitt BDL and Balesdent MH (2005a) Identification and characterisation of polymorphic minisatellites in the phytopathogenic ascomycete *Leptosphaeria maculans*. Current Genetics 47: 37–48

Eckert M, Maguire K, Urban M, Fosters S, Fitt BDL, Lucos JA, Hammond-Kosack KE (2005b) *Agrobacterium tumefaciens* mediated transformation of *Leptosphaeria* spp. and Oculimacula spp. with the reef coral gene *DsRed* and the jellyfish gene gfp. FEMS Microbiology Letters (online: doi: 10.1016 lj.femsle.2005.09.041)

Fernando WGD and Parks PS (2003) First report of blackleg disease caused by *Leptosphaeria maculans* on canola in Brazil. Plant Disease 87: 314

Fitt BDL, Gladders P, Turner JA, Sutherland KG, Welham SJ and Davies JML (1997) Prospects for developing a forecasting scheme to optimise use of fungicides for disease control on winter oilseed rape in the UK. Aspects of Applied Biology 48: 135–142

Gaetan SA (2005) First outbreak of blackleg caused by *Phoma lingam* in commercial canola fields in Argentina. Plant Disease 89: 435

Gladders P and Musa TM (1980) Observations on the epidemiology of *Leptosphaeria maculans* stem canker in winter oilseed rape. Plant Pathology 29: 28–37

Gladders P, Evans N, Marcroft SJ and Pinochet X (2006) Dissemination of information about management strategies and changes in farming practices for the exploitation of resistance to *Leptosphaeria maculans* (phoma stem canker) in oilseed rape cultivars. European Journal of Plant Pathology 114: 117–126

Gudelj I, Fitt BDL and vanden Bosch F (2004) Evolution of sibling fungal pathogens in relation to host specialisation. Phytopathology 94: 789–795

Gugel RK and Petrie GA (1992) History, occurrence, impact and control of blackleg of rapeseed. Canadian Journal of Plant Pathology 14: 36–45

Howlett BJ (2004) Current knowledge of the interaction between *Brassica napus* and *Leptosphaeria maculans*. Canadian Journal of Plant Pathology 26: 245–252

Huang YJ, Fitt BDL and Hall AM (2003a) Survival of A-group and B-group *Leptosphaeria maculans* (phoma stem canker) ascospores and mycelium on oilseed rape stem debris. Annals of Applied Biology 143: 399–369

Huang YJ, Toscano-Underwood C, Fitt BDL, Hu XJ and Hall AM (2003b) Effects of temperature on ascospore germination and penetration of oilseed rape (*Brassica napus*) leaves by A-group or B-group *Leptosphaeria maculans* (phoma stem canker). Plant Pathology 52: 245–253

Huang YJ, Fitt BDL, Jedryczka M, Dakowska S, West JS, Gladders P, Steed JM and Li ZQ (2005) Patterns of ascospore release in relation to phoma stem canker epidemiology in England (*Leptosphaeria maculans*) and Poland (*L. biglobosa*). European Journal of Plant Pathology 111: 263–277

Huang YJ, Li ZQ, Evans N, Rouxel T, Fitt BDL and Balesdent MH (2006) Fitness cost associated with loss of the *AvrLm4* avirulence function in *Leptosphaeria maculans* (phoma stem canker of oilseed rape). European Journal of Plant Pathology 114: 77–89

Jedryczka M, Lewartowska E and Frencel I (1994) Properties of *Phoma lingam* (Tode ex Fr.) Desm. isolates from Poland. I. Pathogenicity characterisation. Phytopathologia Polonica 7: 71–79

Jedryczka M, Nikonorenkov VA, Levitin M, Gasich E, Lewartowska E and Portenko L (2002) Spectrum and severity of fungal diseases on spring oilseed rape in Russia. International Organisation for Biological Control Bulletin 25: 13–20

Karolewski Z, Kosiada T, Hylak-Nowosad B and Nowacka K (2002) Changes in population structure of *Leptosphaeria maculans* in Poland. Phytopathologia Polonica 25: 27–34

Kenyon D, Thomas J and Handy C (2004) Feasibility of using quantitative PCR for assessing resistance to stem canker in oilseed rape cultivars. International Organisation for Biological Control Bulletin 27: 109–117

Khangura RK and Barbetti M (2001) Prevalence of blackleg (*Leptosphaeria maculans*) on canola (*Brassica napus*) in Western Australia. Australian Journal of Experimental Agriculture 41: 71–80

Khangura RK, Barbetti MJ and Walton G (2003) WA blackleg resistance ratings on canola varieties for 2003. Department of Agriculture Western Australia, Farmnote No. 06/2003.

Koch E, Song K, Osborn TC and Williams PH (1991) Relationship between pathogenicity, phylogeny based on restriction fragment length polymorphism in *Leptosphaeria maculans*. Molecular Plant-Microbe Interactions 4: 341–349

Kuhn MH, Gout L, Howlett BJ, Melayah D, Meyer M, Balesdent MH and Rouxel T (2006) Genetic linkage maps and genomic organization in *Leptosphaeria maculans*. European Journal of Plant Pathology 114: 17–31

Laing MD (1986) The crucifer blackleg pathosystem in Natal, South Africa. Acta Horticulturae 194: 141–151

Leach JE, Vera Cruz CM, Bai J and Leung H (2001) Pathogen fitness penalty as a predictor of durability of disease resistance genes. Annual Review of Phytopathology 39: 187–224

Li H, Sivasithamparam K and Barbetti MJ (2003) Breakdown of a *Brassica rapa* ssp. *sylvestris* single dominant blackleg resistance gene in *B. napus* rapeseed by *Leptosphaeria maculans* field isolates in Australia. Plant Disease 87: 752

Li H, Barbetti MJ and Sivasithamparam K (2005) Hazard from reliance on cruciferous hosts as sources of major gene based resistance for managing blackleg (*Leptosphaeria maculans*) disease. Field Crops Research 91: 185–191

Marcroft SJ, Purwantara A, Salisbury PA, Potter TD, Wratten N, Khangura R, Barbetti MJ and Howlett BJ (2002) Reaction of a range of *Brassica* species under Australian conditions to the fungus *Leptosphaeria maculans*, the causal agent of blackleg. Australian Journal of Agricultural Research 42: 587–594

Marcroft SJ, Sprague SJ, Salisbury PA and Howlett BJ (2004) Potential for using host-resistance to reduce production of pseudothecia and ascospores of *Leptosphaeria maculans*, the blackleg pathogen of *Brassica napus*. Plant Pathology 53: 468–474

McDonald B and Linde C (2002) Pathogen population genetics, evolutionary potential and durable resistance. Annual Review of Phytopathology 40: 349–379

Mendes-Pereira E, Balesdent MH, Brun H and Rouxel T (2003) Molecular phylogeny of the *Leptosphaeria maculans–L. biglobosa* species complex. Mycological Research 107: 1287–1304

Moreno-Rico O, Frias-Trevino AG, Luna-Ruiz JJ, Manzano-Flores DE, Romero-Cova S and Seguin-Swartz G (2001) Characterisation and pathogenicity of isolates of *Leptosp-*

haeria maculans from Aguascalientes and Zacatecas, Mexico. Canadian Journal of Plant Pathology 23: 270–278

Piening L, Okolo E and Harder D (1975) Blackleg of rapeseed in Kenya. East African Agriculture and Forestry Journal 41: 110–113

Petrie GA (1995) Long-term survival and sporulation of *Leptosphaeria maculans* (blackleg) on naturally-infected rapeseed/canola stubble in Saskatchewan. Canadian Plant Disease Survey 75: 23–34

Pietravalle S, Lemarie S and van den Bosch F (2006) Durability of resistance and cost of virulence. European Journal of Plant Pathology 114: 107–116

Pilet ML, Duplan G, Archipiano M, Barret P, Baron C, Horvais R, Tanguy X, Lucas MO, Renard M and Delourme R (2001) Stability of QTL for field resistance to blackleg across two genetic backgrounds in oilseed rape. Crop Science 41: 197–205

Plummer KM, Dunse K and Howlett BJ (1994) Non-aggressive strains of the blackleg fungus, *Leptosphaeria maculans*, are present in Australia and can be distinguished from aggressive strains by molecular analysis. Australian Journal of Botany 42: 1–8

Rimmer SR and van den Berg CGJ (1992) Resistance of oilseed *Brassica* spp. to blackleg caused by *Leptosphaeria maculans*. Canadian Journal of Plant Pathology 14: 56–66

Rouxel T and Balesdent MH (2005) The stem canker (blackleg) fungus, *Leptosphaeria maculans*, enters the genomic era. Molecular Plant Pathology 6: 225–241

Rouxel T, Willner E, Coudard L and Balesdent MH (2003a) Screening and identification of resistance to *Leptosphaeria maculans* (stem canker) in *Brassica napus* accessions. Euphytica 133: 219–231

Rouxel T, Penaud A, Pinochet X, Brun H, Gout L, Delourme R, Schmit J and Balesdent MH (2003b) A 10-year survey of populations of *Leptosphaeria maculans* in France indicates a rapid adaptation towards the *Rlm1* resistance gene of oilseed rape. European Journal of Plant Pathology 109: 871–881

Salam MU, Khangura RK, Diggle AJ and Barbetti MJ (2003) Blackleg Sporacle: a model for predicting onset of pseudothecia maturity and seasonal ascospore showers in relation to blackleg in canola. Phytopathology 93: 1073–1081

Salisbury PA, Ballinger DJ, Wratten N, Plummer KM and Howlett BJ (1995) Blackleg disease on oilseed Brassica in Australia: a review. Australian Journal of Experimental Agriculture 35: 665–672

Sexton AC and Howlett BJ (2001) Green fluorescent protein as a reporter in the Brassica–*Leptosphaeria maculans* interaction. Physiological and Molecular Plant Pathology 58: 13–21

Shoemaker RA and Brun H (2001) The teleomorph of the weakly aggressive segregate of *Leptosphaeria maculans*. Canadian Journal of Botany 79: 412–419

Somda I, Harkous S and Brun H (1997) Bipolar heterothalism in B-group isolates of *Leptosphaeria maculans*. Plant Pathology 46: 890–896

Somda I, Renard M and Brun H (1998) Seedling and adult plant reactions of *Brassica napus–B. juncea* recombinant lines towards A- and B-group isolates of *Leptosphaeria maculans*. Annals of Applied Biology 132: 187–196

Sprague SJ, Balesdent MH, Brun H, Hayden HL, Marcroft SJ, Pinochet X, Rouxel T and Howlett BJ (2006) Major gene resistance in *Brassica napus* (oilseed rape) is overcome by changes in virulence of populations of *Leptosphaeria maculans* in France and Australia. European Journal of Plant Pathology 114: 33–40

Szlávik SZ, Jedryczka M, Kiss I, Lewartowska E and Nagy G (2003) Population structure and pathogenicity grouping of *L. maculans* isolates from Hungary. Blackleg News 3–4

Thomas J and Wedgwood E (1998) Potential for exploiting resistance to stem canker (*Leptosphaeria maculans*) in cultivars of winter oilseed rape. International Organisation for Biological Control Bulletin 21: 91–96

Toscano-Underwood C, West JS, Fitt BDL, Todd AD and Jedryczka M (2001) Development of phoma lesions on oilseed rape leaves inoculated with ascospores of A-group or B-group *Leptosphaeria maculans* (stem canker) at different temperatures and wetness durations. Plant Pathology 50: 28–41

Toscano-Underwood C, Huang YJ, Fitt BDL and Hall AM (2003) Effects of temperature on maturation of pseudothecia of *Leptosphaeria maculans* and *L. biglobosa* on oilseed rape stem debris. Plant Pathology 52: 726–736

van den Bosch F and Gilligan CA (2003) Measures of durability of resistance. Phytopathology 93: 616–625

Welham SJ, Turner JA, Gladders P, Fitt BDL, Evans N and Baierl A (2004) Predicting light leaf spot (*Pyrenopeziza brassicae*) risk on winter oilseed rape (*Brassica napus*) in England and Wales, using survey, weather and crop information. Plant Pathology 53: 713–724

West JS, Biddulph JE, Fitt BDL and Gladders P (1999) Epidemiology of *Leptosphaeria maculans* in relation to forecasting stem canker severity on winter oilseed rape in the UK. Annals of Applied Biology 135: 535–546

West JS, Evans N, Liu S, Hu B and Peng L (2000) *Leptosphaeria maculans* causing stem canker of oilseed rape in China. Plant Pathology 49: 800

West JS, Kharbanda P, Barbetti MJ and Fitt BDL (2001) Epidemiology and management of *Leptosphaeria maculans* (phoma stem canker) in Australia, Canada and Europe. Plant Pathology 50: 10–27

West JS, Balesdent MH, Rouxel T, Nancy JP, Huang YJ, Roux J, Steed JM, Fitt BDL and Schmit J (2002a) Colonisation of winter oilseed rape tissues by A/Tox^{+} and B/Tox^{0} *Leptosphaeria maculans* (phoma stem canker) in France and England. Plant Pathology 51: 311–321

West JS, Fitt BDL, Leech PK, Biddulph JE, Huang YJ and Balesdent MH (2002b) Effects of timing of *Leptosphaeria maculans* ascospore release and fungicide regime on phoma leaf spot and phoma stem canker development on winter oilseed rape (*Brassica napus*) in southern England. Plant Pathology 51: 454–463

Wherrett AD, Sivasithamparam K and Barbetti MJ (2004) Establishing the relationship of ascospore loads with blackleg (*Leptosphaeria maculans*) severity on canola (*Brassica napus*). Australian Journal of Agricultural Research 55: 849–854

Williams RH and Fitt BDL (1999) Differentiating A and B groups of *Leptosphaeria maculans*, causal agent of stem canker (blackleg) of oilseed rape. Plant Pathology 48: 161–175

Zhou Y, Fitt BDL, Welham SJ, Gladders P, Sansford CE and West JS (1999) Effects of severity and timing of stem canker (*Leptosphaeria maculans*) symptoms on yield of winter oilseed rape (*Brassica napus*) in the UK. European Journal of Plant Pathology 105: 715–728

European Journal of Plant Pathology (2006) 114: 17–31
DOI 10.1007/s10658-005-3168-6

Genetic linkage maps and genomic organization in *Leptosphaeria maculans*

Marie-Line Kuhn[1], Lilian Gout[1,2], Barbara J. Howlett[3], Delphine Melayah[1,4], Michel Meyer[1], Marie-Hélène Balesdent[1] and Thierry Rouxel[1,*]
[1]*INRA-PMDV, Route de St Cyr, 78026 , Versailles Cedex, France;* [2]*Protection des Plantes, INA-PG, F-78850, Thiverval-Grignon, France;* [3]*School of Botany, The University of Melbourne, 3010, Victoria, Australia;* [4]*Laboratoire d'Ecologie Microbienne, Université Claude Bernard Lyon 1, 69622, Villeurbane cedex, France; *Author for correspondence (Phone: +33-130833229; Fax: +33-130833195; E-mail: rouxel@versailles.inra.fr)*

Accepted 6 September 2005

Key words: chromosome length polymorphism, genetic map, genome structure, physical map, repetitive DNA, retrotransposons

Abstract

Leptosphaeria maculans is a haploid outcrossing ascomycete with a genome size of about 34 Megabases (Mb) which is predicted to have between 10,000 and 12,000 genes. The chromosomes of *L. maculans* are of a size range (0.7–3.5 Mb) and number (15–16) that can be readily resolved by pulsed field gel electrophoresis. Chromosome length polymorphisms are generated by meiosis giving rise to size differences as high as 57%, in the case of the ribosomal DNA-harbouring chromosome whose size varies between 1.8 and 4.2 Mb. Genetic maps are characterised by linkage groups comprising an accumulation of markers based on retrotransposon sequences. This, along with sequencing of pericentromeric regions and stretches of ORF-rich regions, suggest that the genome of *L. maculans* consists of a mosaic of GC-equilibrated coding regions with no or few transposons, and of stretches of highly degenerated and truncated retrotransposons, encompassing very few genes. Chromosome length polymorphisms are linked with the loss of dispensable regions. We suggest that the degree of length polymorphism for a particular chromosome correlates to the amount of dispensable retrotransposons, and that some gene-rich chromosomes may be less prone to undergo chromosome length polymorphisms than other chromosomes.

Abbreviations: AFLP – amplified fragment length polymorphism; BAC – bacterial artificial chromosome; CHEF – contour-clamped homogeneous electric field; CLP – chromosome length polymorphism; ISSR – inter-simple sequence repeat, LG – linkage group; LTR – long terminal repeats; RAPD – random amplified polymorphic DNA; Rep-PCR – repetitive element-based PCR; RIP – repeat induced point mutation; TAFE – transverse alternating field electrophoresis

Introduction

Leptosphaeria maculans is an ascomycete fungus, which has historically been classified as a Loculoascomycete, a class comprising over 6000 species (Silva-Hanlin and Hanlin, 1999). However, taxonomic revisions now assign *L. maculans* to the largest order of the Dothideomycetes class (only partly corresponding to the Loculoascomycete class), the Pleosporales (Liew et al., 2000). Pleosporales also encompass other important plant pathogens such as the closely related *Phaeosphaeria (Stagonospora) nodorum* (formerly known as *Leptosphaeria nodorum)*, and other genera such as *Cochliobolus*, *Pleospora*, *Alternaria*, *Venturia* and *Pyrenophora* (Berbee, 2001).

Due to the incidence of phoma stem canker worldwide (Fitt et al., 2005) and the experimental tractability of *L. maculans*, major efforts in understanding the epidemiology and the population genetics of the fungus, and its interaction with the host have been recently undertaken (Balesdent et al., 2002; Delourme et al., 2004; Howlett, 2004; Aubertot et al., 2005; Delourme et al., 2005; Fitt et al., 2005). In addition *L. maculans* exemplifies Dothideomycete infection strategies and has characteristics in its life traits, including pathogenicity, that render it an excellent model to address numerous issues related to pathogen evolution, pathogenicity and host specificity. The importance of sexual recombination in the wild also renders *L. maculans* an appropriate model for studying molecular evolution of genes submitted to selection pressure (avirulence genes, fungicide resistance genes) (Rouxel and Balesdent, 2005).

Numerous tools have been developed that will allow the *L. maculans* scientific community to enter the 'genomic era' through the use of a gene-by-gene approach. These include efficient transformation techniques, large collections of segregating populations (including tetrads), genetic maps, Bacterial Artificial Chromosomes (BACs) and cosmid libraries, Expressed Sequence Tag libraries, including subtractive libraries of the plant pathogen interaction, and collections of strains tagged with selectable markers that can mutate genes (for a review see Rouxel and Balesdent, 2005). However, to date, like other Dothideomycetes, the genome structure of *L. maculans* is poorly characterised. The genome size of the fungus is about 34 Megabases (Mb) and this is predicted to contain between 10,000 and 12,000 genes (Cozijnsen et al., 2000). The chromosomes of *L. maculans* are of a size range (0.7–3.5 Mb) and number (15–16) that can be readily resolved by pulsed field gel electrophoresis (Howlett et al., 2001).

This paper presents our current knowledge on the genome of *L. maculans* focusing on structural features that can be deduced from (i) genetic maps (ii) physical maps built by hybridisation of probes to electrokaryotypes, and (iii) sequence data on clusters of coding regions or repeats. Our data strongly suggest that, at least for some chromosomes, the genome *of L. maculans* encompasses mosaics of large coding regions and large dispensable zones, which may be responsible for chromosome length polymorphisms generated following meiosis.

Materials and methods

Fungal isolates and in vitro crosses

Three crosses were used to build genetic maps at INRA, Versailles. Cross #11 between two French field isolates (a.2 and H5) heterozygous at the avirulence *AvrLm1* locus, and cross #23 between two isolates (v11.3.6, a progeny of the a.2 × H5 *in vitro* cross, and 21.3.1, a progeny of a cross between a German field isolate and a French lab. isolate) heterozygous at the *AvrLm1* and *AvrLm4* loci, were described by Attard et al. (2002). Cross #30 between v23.1.2 (a progeny of the v11.3.6 × 21.3.1 cross displaying the *AvrLm7* allele) and the New Zealand field isolate Nzt 4 (virulent for this locus) was described by Balesdent et al. (2002). Maps 11, 23 and 30 were constructed from 98, 42 and 46 F_1 progeny, respectively. Isolates were cultured and maintained as previously described (Ansan-Melayah et al., 1995; Attard et al., 2002).

Markers used to generate the genetic maps

Except for biological (mating type) and phytopathological (host-specificity, i.e., avirulence) loci, markers were mostly anonymous and PCR-based (Table 1). These encompass RAPD (Ansan-Melayah et al., 1995), ISSR, e.g., markers designed to amplify between microsatellites (West et al., 2002), AFLP, and Rep-PCR markers (Jedryczka et al., 1999), i.e., mostly markers generated by using the sequence polymorphism of the Long Terminal Repeat (LTR) retrotransposon *Pholy* (Attard et al., 2005) to amplify between tandem truncated copies of *Pholy* (Jedryczka et al., 1999). Protocols for generation, separation and visualization of RAPD, ISSR and Rep-PCR are described in Ansan-Melayah et al. (1995), West et al. (2002), and Jedryczka et al. (1999), respectively.

AFLP analysis was performed according to Vos et al. (1995) with slight modifications and amplified products run on acrylamide gels. Genomic DNA (approx. 500 ng) from parents and F_1 progeny was digested with 5 U of *Mse*I (Invitrogen) (30 min digestion at 37 °C), followed by a digestion with 10 U of *Eco*RI (Invitrogen) (overnight digestion at 37 °C). Adaptors were designed and ligated to the restriction fragments according

Table 1. Characteristics of three genetic maps of *Leptosphaeria maculans* developed at INRA, Versailles

Cross number	11	23	30
Total number of markers in the map, including	443	177	266
RAPD	102	6	4
ISSR	48	6	29
Rep-PCR	35	0	1
Mini- and microsatellites	3	0	2
AFLP	235	163	229
Others	20	2	1
Major linkage groups			
More than 20 cM	41 (3–13 markers)	7 (4–9 markers)	7 (4–9 markers)
More than 6 markers	20	4	7
Doublets	17	19	27
Unlinked markers	76 (17.1%)	60 (33.9%)	54 (20.3%)
Total number of linkage groups	78	35	61
Map coverage (cM)	2129	430	867
Average physical/genetic distance (kb/cM) [a]	16	79	39

[a]As estimated on the basis of a 34 Mb genome size (Cozijnsen et al., 2000).

to Vos et al. (1995). A 3 μl part of the mix was first amplified with a +1-*Mse*I primer and a +1-*Eco*RI primer, i.e., primers with one additional selective nucleotide, followed by a second round of selective amplification using *Eco*RI + 1 and *Mse*I + 2 primer combination. In accordance with findings of Cozjinsen et al. (2000) this combination gave rise to more resolvable bands per lane than the other combinations and consequently more polymorphic bands between parents. Amplification took place in a 20 μl volume with 1 μM of each primer, 0.1 μM of each dNTP, 2 mM $MgC1_2$, 2 μl 10× Taq buffer and 1.25 U Taq Polymerase (Applied Biosystems). Amplification took place in a PE-9600 Thermal Cycler (Applied Biosystems). Cycling conditions were as follows: first round of amplification: cycle 1–20; 1 s at 94 °C, 30 s at 56 °C, 2 min at 72 °C, with 0.01 ramp. Selective amplification: cycle 1; 2 min at 94 °C, 30 s at 65 °C, 2 min at 72 °C. Cycle 2–9; 1 s at 94 °C, 30 s at 64 °C with a progressive drop of 1 °C annealing temperature in each cycle, 2 min at 72 °C. Cycle 10–32; 1 s at 94 °C, 30 s at 56 °C, 2 min at 72 °C. Two microlitres of the amplification products were separated in a CastAway Precast 5.5% Long Ranger acrylamide gel (Stratagene) and electrophoresed in a CastAway Sequencing Device (Stratagene) according to the manufacturer's recommendations. After electrophoresis the gel was stained with silver nitrate as described by Chalhoub et al. (1997).

Genetic map construction

Polymorphic bands were recorded as present/absent in the progeny and monogenic segregation was checked with a Chi-squared test ($P = 0.05$). Monogenic markers were submitted to multipoint linkage analysis using the Mapmaker/Exp 3.0 software (Lander et al., 1987) set for F_2 backcross segregating population, with log of the likelihood ratio (LOD) value of 3.0, a maximum distance of 20 cM. Linkage groups were drawn with the Mappit version 1.3 software (L. Gianfranschi and B. Koller, Swiss Federal Institute of Technology, http://www.pa. ipw.agrl.ethz.ch/).

Cloning and characterization of markers

AFLP bands were isolated from the acrylamide gels as described by Chalhoub et al. (1997) and fragments ligated into vector pGEM-T Easy (Promega). Cloning and sequencing was as described previously (Attard et al., 2001). Minisatellites were obtained as described by Eckert et al. (2005). Chromosomal DNA was prepared according to Plummer and Howlett (1995) and resolved by contour-clamped homogeneous electric fields (CHEF) electrophoresis using the conditions described by Morales et al. (1993). Chromosomal DNA was vacuum blotted on nylon N^+ membranes and hybridized as described previously (Leclair et al., 1996).

Results and discussion

Genetic linkage maps of L.maculans

Genetic linkage maps have been developed for *L. maculans* by Pongam et al. (1998), Cozijnsen et al. 2000) and by the INRA-PMDV group (this study). The more detailed one is presented in Figure 1. The map developed by Pongam et al. (1998) comprised 56 AFLP markers on 67 progeny from a cross between the Australian isolate PHW1223 and the French isolate PHW1245 (IBCN74) that segregated for an avirulence gene that the authors named *alm1*. This gene probably corresponds to *AvrLm4* (Delourme et al., 2004). Nine linkage groups and five pairs of markers were assigned with a total genome size of 340 cM and an average distance between loci of 6.1 cM. Cozijnsen et al. (2000) constructed a genetic map comprising 155 AFLP markers, three RAPD markers, the mating

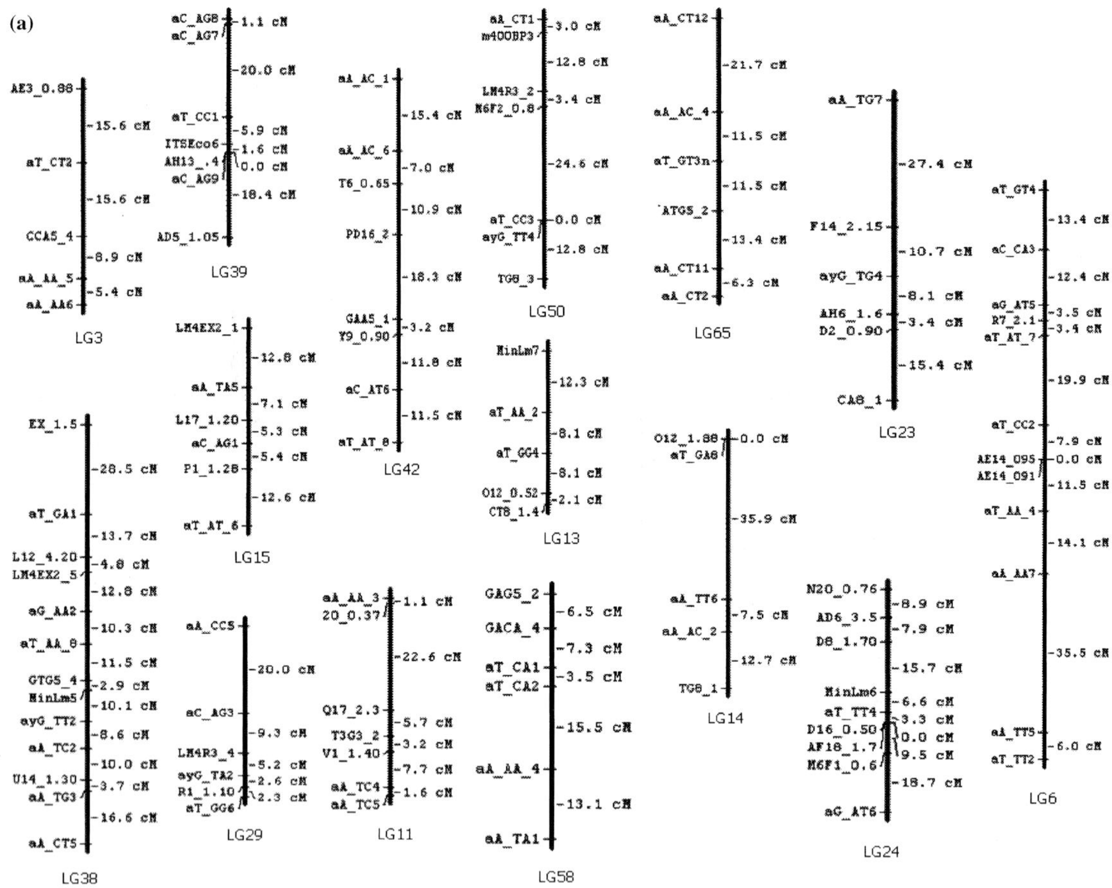

Figure 1. The linkage map of *Leptosphaeria maculans* cross #11. (a) shows the major linkage groups (LG) (5 markers or more, and more than 30 cM) (b) shows LG harbouring 5 markers or more, with accumulation of markers over a small genetic distance. The map was derived by analysis with Mapmaker/Exp 3.0 software set for F_2 backcross segregating population (LOD = 3.0, maximum distance = 20 cM). The marker names are shown to the left of the LG and the distance between markers (cM) is indicated on the right of the LG. AFLP markers are indicated with an 'a' followed by the additional 3' nucleotides of the primer combination (e.g., A_TA) and the number of the polymorphic band on the AFLP gel. ISSR markers are indicated as the name of the repeat used as primer (e.g., TG8 or GAG5) followed by the number of the polymorphic band on the agarose gel. RAPD markers are indicated as the name of the primer (as per Operon Technology; e.g., L12, U14 or AH13) followed by the size (kb) of the polymorphic band. Rep-PCR markers are indicated in the FxRy form (some are LmxEXy or LmxRy) followed by the size (kb) of the marker band. MAT denotes the mating type locus, AvrLmx corresponds to avirulence loci, ITSEco6 to the rDNA locus, and MinLmx to minisatellite loci.

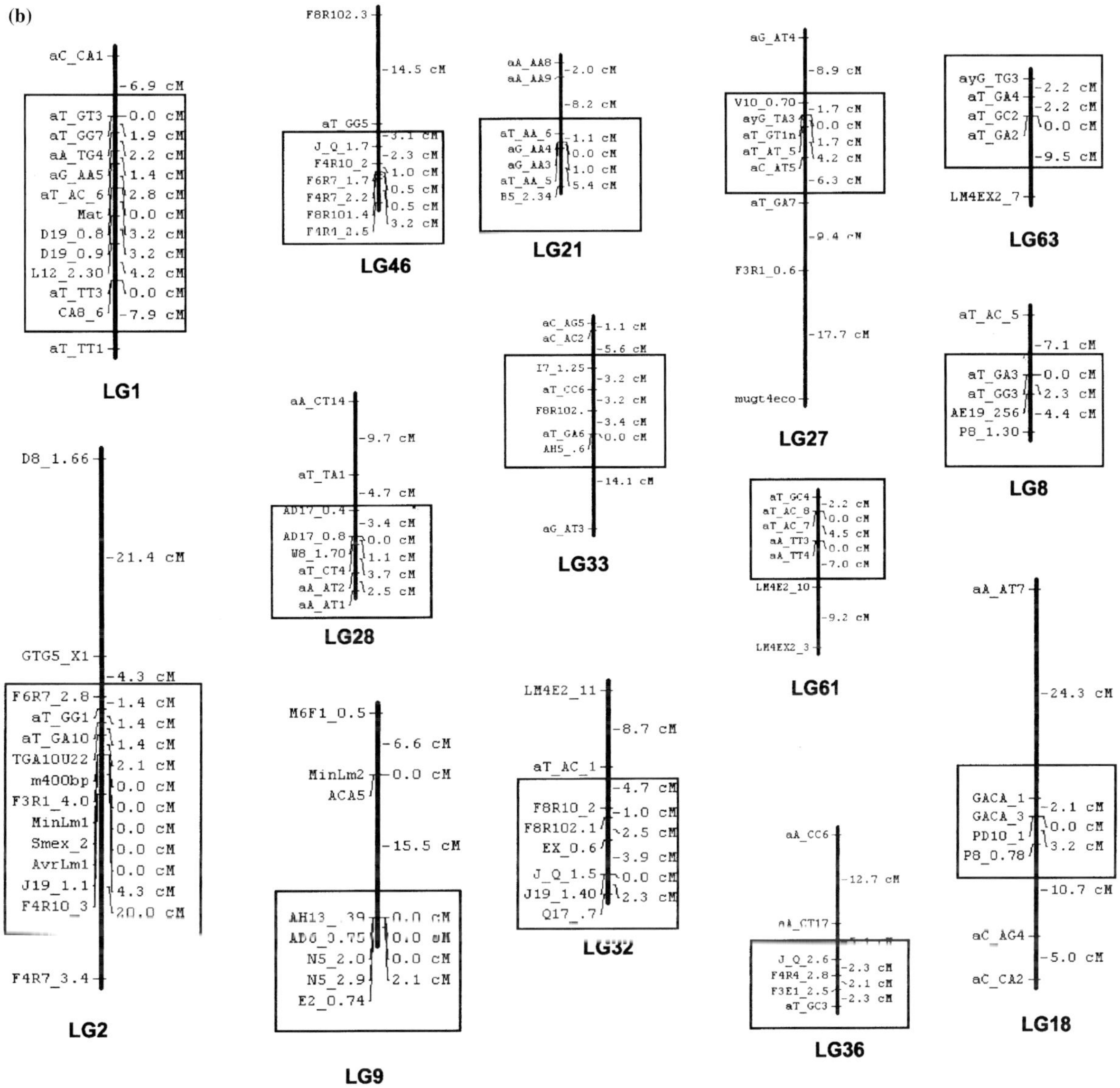

Figure 1. (Continued).

type locus, and a *B. juncea* host specificity locus. The cross was between Australian isolates C13 and Ml (IBCN17 and 18, respectively) (Figure 2) and the mapping was performed on 58 F_1 progeny Twenty-one linkage groups, five pairs, and 18 unlinked markers were assigned, and the coverage was 1520 cM.

The *L. maculans* INRA maps comprised between 177 and 443 markers, mainly including anonymous PCR-based markers (Table 1). The most detailed map (from cross #11) was obtained using a F_1 segregating population of 98 progeny. This map had 15 major groups (comprising 5 markers or more, and covering more than 30 cM) (Figure 1a). Its overall coverage (2129 cM), and physical to genetic ratio is comparable to the most detailed linkage maps currently available for other fungal species (Table 2). However, 14 additional groups that could be considered as major groups in terms of number of markers, were characterised by an accumulation of markers in a very small genetic distance (Figure 1b). This was also the case for the less detailed genetic maps developed from crosses #23 and #30, with each map showing seven linkage groups again with several markers in a small genetic distance (data not shown). The mating-type locus, *MAT*, and the avirulence *AvrLm1* locus mapped to these linkage groups (Figure 1b).

Our data indicate that the other two avirulence loci *AvrLm4* and *AvrLm7* mapped in crosses #23 and #30 are also in a region with many markers in a small distance (Figure 3). Such a location of avirulence/virulence loci in linkage groups with many markers in a small genetic distance, and/or clustering of avirulence loci so that they form a separated linkage group, seems to be a usual feature of fungal genetic maps, as noted for *Blumeria graminis* (Pedersen et al., 2002) and for the two Dothideomycetes, *Mycosphaerella graminicola* (Kema et al., 2002) and *Cochliobolus sativus* (Zhong et al., 2002). In addition, for our *L. maculans* maps, there was a difference between these two types of linkage groups in terms of the major markers that defined them. Groups that had many markers over a small genetic distance were rich in Rep-PCR markers, i.e. markers based on the *Pholy* retrotransposon (23 Rep-PCR markers out of 113), whereas these markers were almost absent from the major linkage groups displayed on Figure 1a (5 markers out of 108). The opposite situation was observed for ISSR markers (10.2% in the major groups compared to 4.4% in groups showing accumulation of markers) (Figure 1). This difference was not observed for AFLP

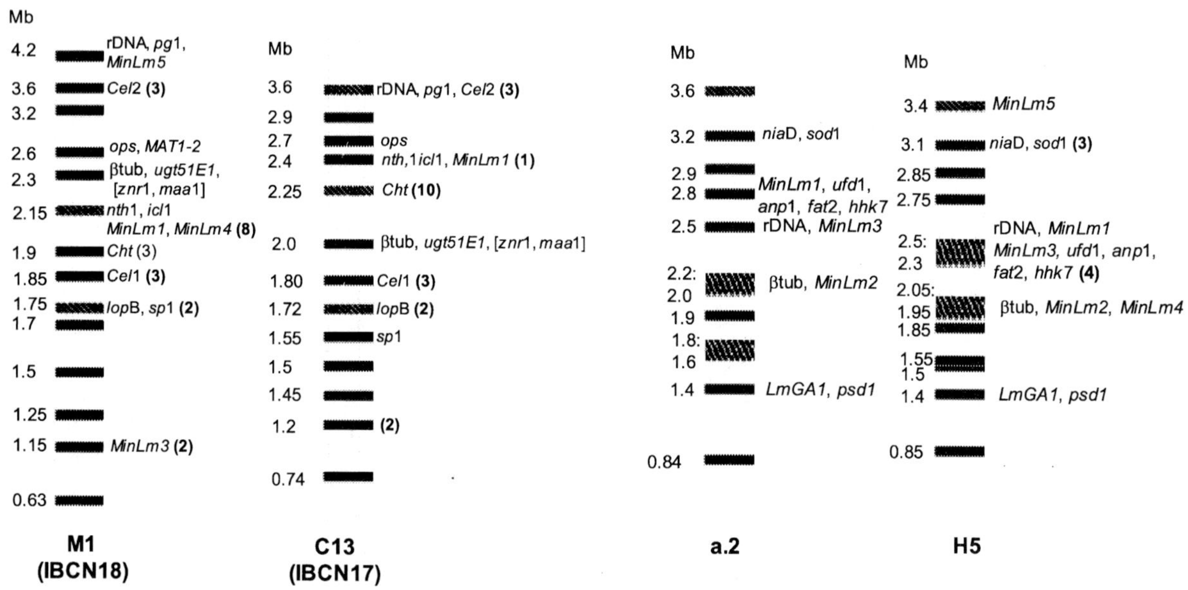

Figure 2. Schematic diagram of chromosomes of isolates M1, C13, a.2 and H5 showing estimated sizes (Mb) and location of probes corresponding to identified genes or minisatellites of *L. maculans*. The pattern is a synthetic representation based on separations done on gels run under several conditions to resolve the whole karyotype. In addition, the number of cloned markers, originating from the genetic maps and allocated to a given chromosome is indicated in bold in parentheses. Bands that stain brightly with ethidium bromide (indicated by diagonal striping) probably comprise more than one chromosome. The smallest chromosome (sized 0.63–0.85 Mb) is thought to be a dispensable 'B' chromosome (Leclair et al., 1996), with no single-copy sequence detected to date. Probes used for hybridisation are: *MinLm1-5*, minisatellite loci (Eckert et al., 2005); *LmGA1*, G protein alpha subunit, physically linked to *psd1*, phosphatidyl serine decarboxylase; *sod1*, superoxide dismutase 1; *ufd1*, ubiquitin fusion degradation protein 1; *anp1*, similar to ANP1 protein; *fat2*, fatty acid transporter 2; *hhk7*, histidine kinase 7 (this study); βtub, β-tubulin (Cozijnsen et al., 2000; this study); rDNA, ribosomal DNA (Howlett et al., 1997; this study); *niaD*, nitrate reductase, physically linked to *niiA*, nitrite reductase (Williams et al., 1994, 1995; this study); *pg1;* polygalacturonase 1; *Cel1*, cellulase 1; *Cel2*, cellulase 2 (Sexton et al., 2000)*;* *ops*, opsin (Idnurm and Howlett, 2001); *MAT1–2*, mating-type locus, physically linked to *GAP*, GTPase activating protein, and *DNA-L*, a DNA-Lyase homologue (Cozijnsen and Howlett, 2003); *icl1*, isocitrate lyase physically linked to *prs5*, ribose phosphate pyrophosphokinase (Idnurm and Howlett, 2002); *znr1* and *maa1*, physically linked to multidrug facilitator 1 (*mfs1*), elongation factor (*ef1b*), heat shock protein (*hsp78*), DNA mismatch repair protein (*msh5*) (Idnurm et al., 2003a); *nth1*, neutral trehalase 1, physically linked to *ade3*, C1 tetrahydrofolate synthase; *ugt51E1*, UDP-glucose:sterol glucosyltransferase, physically linked to MAP kinase (Idnurm et al., 2003b); *Cht*, cyanide hydratase (Sexton and Howlett, 2000); *lop-B*, loss of pathogenicity B (Idnurm and Howlett, 2003); *sp1*, secreted protein 1, physically linked to *chp1*, conserved hypothetical protein (Wilson et al., 2002). Minisatellites *MinLm1*, *MinLm2*, and *MinLm5* map to LG2, LG9 and LG38 of the cross #11 genetic map, respectively (see Figure 1).

Table 2. General features of the genetic linkage maps constructed for fungal species using anonymous molecular marker[a]

Fungal species	Progeny[b]	Markers[b]	Type of Markers[c]	Chromosomes	Major LG	Accumulation of markers in LGs	Physical map[d]	CLP[e]	Map coverage (cM)	Physical/genetic Ratio (kb/cM)	References[f]
Basidiomycetes											
Cryptococcus neoformans var *neoformans*	100	181	AFLP/RAPD	13	14	Limited	Yes	Yes	1917	13.6	(1)
	94	289	RFLP/SSR	14	20	Yes	Yes	Yes	1536	13.2	(2)
Coprinus cinereus	40	256	RAPD/RFLP	13	13	Yes	Yes	Yes	1346	27.9	(3)
Pleurotus ostreatus	80	189	RAPD/RFLP	11	11	No	Yes	Yes	1001	35.1	(4)
Ascomycetes: Sordariomycetes											
Magnaporthe grisea	61	254	RFLP/Rep/SSR	7	7	Yes	Yes	?	900	44.4	(5,6)
Gibberella zeae	99	1052	AFLP	4	9	Yes	No	?	1300	30.7	(7,8)
Gibberella moniliformis	121	636	AFLP/RFLP	12	12	Yes	Yes	?	2188	21	(9)
Ascomycetes: Leotiomycetes											
Blumeria graminis	81	359	AFLP/RFLP/SNP/Rep	7 or 8	34	Yes	No	?	2114	*c.* 19	(10)
Ascomycetes: Dothideomycetes											
Mycosphaerella graminicola	61	282	AFLP/RAPD	17–18	23	Yes	Limited	Yes	1216	28	(11)
Cochliobolus heterostrophus	91	120	RFLP	15–16	26	No	Yes	Yes	1505	23	(12)
Cochliobolus sativus	104	134	AFLP/RFLP	15	27	Yes	Yes	Yes	1329	25	(13)

[a]'Founding' fungal species whose extremely saturated maps encompass hundreds of genes initially identified by phenotypes are not included in this Table. These include *Saccharomyces cerevisiae*, *Neurospora crassa* and to a lesser extent *Aspergillus nidulans*.
Information and links to the genetic maps of these species are at http://www.broad.mit.edu/annotation/fungi/and http://www.yeastgenome.org/

[b]Number of progeny and markers used to construct the genetic map.

[c]Major types of markers used to generate the map.

[d]Hybridization of markers to electrokaryotypes and assignment of LG to given chromosomes.

[e]Existence of large Chromosome Length Polymorphisms (CLPs) in the fungal species.

[f]References, (1) Forche et al. (2000); (2) Marra et al. (2004); (3) Muraguchi et al. (2003); (4) Larraya et al. (2000); (5) Nitta et al. (1997); (6) http://www.broad.mit.edu/annotation/fungi/magnaporthe/; (7) Jurgenson et al. (2002a); (8) description of another *G. zeae* map is at http://www.broad.mit.edu/annotation/fungi/fusarium; (9) Jurgenson et al. (2002b); (10) Pedersen et al. (2002); (11) Kema et al. (2002), (12) Tzeng et al. (1992); (13) Zhong et al. (2002).

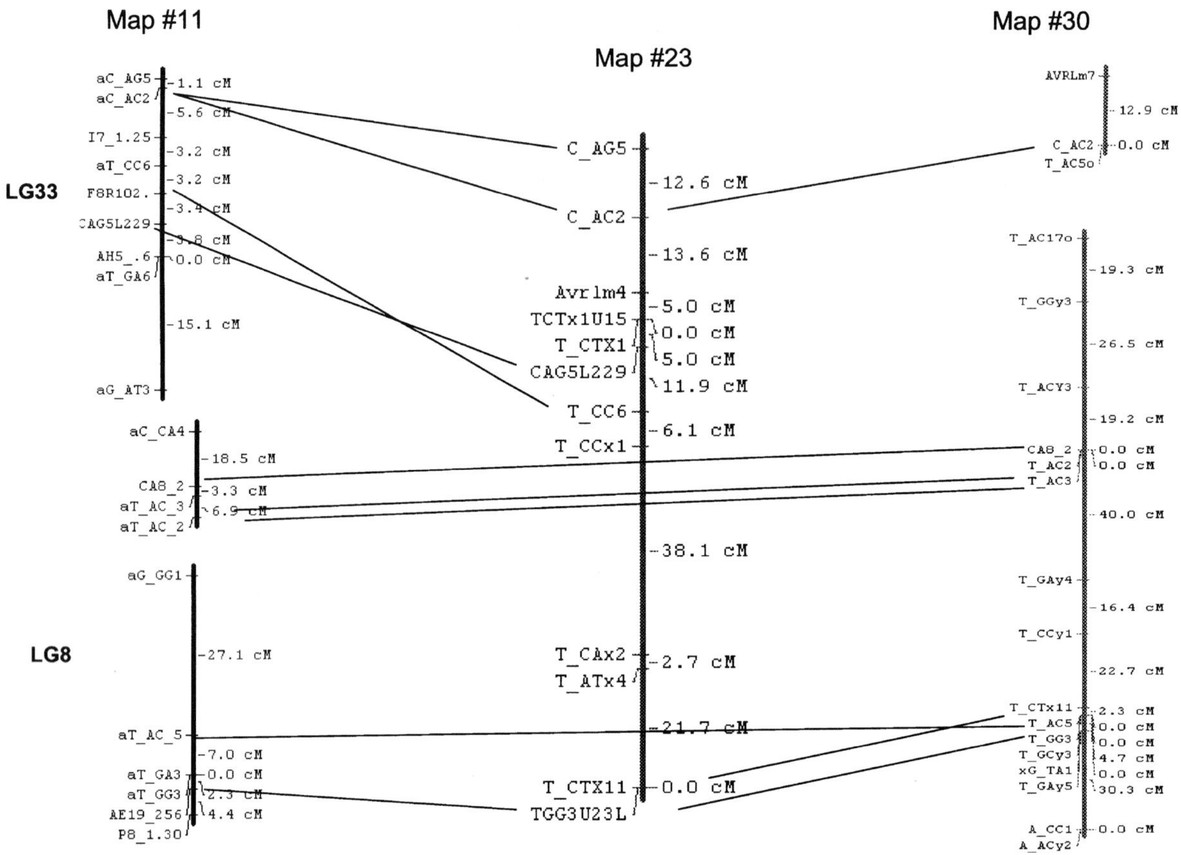

Figure 3. An example of co-linearization between the three *L. maculans* maps developed at INRA. The maps were constructed with Mapmaker/Exp 3.0 software set for F_2 backcross segregating population (LOD = 3.0, maximum distance = 30 cM). Common AFLP markers between the three maps are indicated with a plain line. Marker name nomenclature and distances are as in Figure 1, except that names of AFLP markers do not begin with a lowercase 'a' in maps #23 and #30.

markers (53.7% in the major groups compared to 50% in groups showing accumulation of markers) or for RAPD markers (24.1% in the major groups compared to 23% in groups showing accumulation of markers) (Figure 1). Again, accumulation of specific types of markers, particularly anonymous or retrotransposon-based markers, is commonly observed in fungal genetic maps (Table 2). For example, retrotransposons such as those of the MAGGY Gypsy-type are clustered at specific locations in the *Magnaporthe grisea* integrated genetic map as are other types of repetitive DNA (Nitta et al., 1997). However, markers based on repeats are still uncommon in fungal genetic maps and clustering of markers is often attributed to the use of AFLP markers whereas RAPD, and mainly RFLP markers or microsatellites are less prone to such clustering (e.g. Kema et al., 2002; Jurgenson et al., 2002a, 2002b; Zhong et al., 2002). As emphasized by Jurgenson et al. (2002a), the most likely explanation for clustering of AFLP markers is the map saturation that can be attained with such markers. Accordingly, the relatively low number of AFLP markers used here in the *L. maculans* genetic map could explain the lack of specific clustering of these markers. Other explanations for clustering of such markers suggested by Jurgenson et al. (2002a) include non-random distribution of AT-rich regions in the genome, which would contain a higher number of *Eco*RI and *Mse*I restriction sites, and/or recombination suppression in some genomic regions.

AFLP markers common between the three INRA maps were investigated and cloned in order to co-linearize the three maps. Forty-three markers out of 98 tested (43.8%) common between map #11 and map #23 were obtained; 64 common markers out of 187 tested (34.2%) were obtained between map #11 and map #30; and 16 markers out of 80 tested (20.0%) were common between map #23 and map #30. This latter low number of common markers between map #23 and map #30 led to only eight markers being common to the three maps. Using standard MapMaker settings (LOD 3.0, max distance 20 cM), the co-linearization strategy was unable to combine linkage groups. When maximum distance was increased to 30 cM, several previously unlinked groups were combined for some maps (Figure 3). This extensive mapping showed that, even though the markers were scattered throughout the genome, increasing the number of markers usually increased the size of given linkage groups without reducing their number. Accumulation of markers in some linkage groups further suggested that Rep-PCR, ISSR, and to a lesser extent that AFLP or RAPD markers, could specifically target some regions of the genome whereas other genomic regions are extremely poor targets for these markers, therefore leading to gaps in the maps, and to many more linkage groups than chromosomes. This can also be illustrated by similar findings in the linkage map of the Dothideomycete *C. sativus*, with chromosomes encompassing up to three linkage groups, one of which is characterized by accumulation of markers in a small genetic distance (Zhong et al., 2002). In this respect, the recent development of new types of markers, i.e., locus-specific minisatellites, were able to target parts of the genome not covered by other markers, as exemplified by the *MinLm6* minisatellite which allowed two previously unlinked minor LGs to be joined (LG24, Figure 1a).

Cloning of markers and their assignment to physical maps

Cozijnsen et al. (2000) cloned three polymorphic RAPD bands and 18 polymorphic AFLP bands of their map to provide anchoring markers for chromosomes. These AFLP and RAPD markers were hybridised to Southern blots of chromosomal DNA of parent isolates, but only the three RAPD markers and eight of the AFLP markers were chromosome-specific. The remaining 10 AFLP markers bound to all chromosomes. Eight of these were sequenced; two (GenBank AF293766 and AF293768) corresponded to portions of the *Copia* LTR retrotransposon *Pholy* (previously named LMR1; Cozijnsen et al., 2000) whereas the six other markers did not have any match in the database until recently. Here, we have shown that three of these (6 unknown AFLP markers) (GenBank AF293767, AF293769 and AF293770) correspond to various parts of the recently identified *Gypsy*-like LTR retrotransposon *Polly*, and another one (GenBank AF293771) correspond to part of the recently identified *Gypsy*-like LTR retrotransposon *Rolly* (Gout, 2005). Single-copy minisatellites identified in *L. maculans* by Eckert et al. (2005) were also assigned to chromosomes. A schematic representation of the chromosomes of the two parents of the Cozijnsen et al. (2000) map (of isolates M1 and C13) and of map #11 of the present study (of isolates a.2 and H5) is shown in Figure 2, where the letters and numbers next to each band refer to chromosome-specific markers, corresponding to known genes or to minisatellites. The number of anonymous RAPD or AFLP markers per chromosome is also indicated. Most chromosomes have several markers. In contrast, as exemplified by the most extensive physical map available, that of isolate M1 (Figure 2), six (or seven, if including the minichromosome) chromosomes do not possess any marker from the genetic maps. However, as mentioned above, many anonymous RAPD and AFLP markers tested bound to all chromosomes.

Chromosome length polymorphisms

Large variation in chromosome sizes is a feature of many fungal electrokaryotypes (Table 2). Chromosome Length Polymorphisms (CLPs) were observed between natural isolates of *L. maculans* firstly using TAFE (transverse alternating field electrophoresis) and then CHEF (contour-clamped homogeneous electric fields) separation of chromosomal DNA (Taylor et al., 1991; Morales et al., 1993). Morales et al. (1993) also hybridised chromosomal DNA with conserved heterologous sequences, such as β-tubulin, and thus suggested that homologous chromosomes may have very different sizes. As for other Dothideomycetes species (Tzeng et al., 1992; Zhong et al., 2002; Kema et al., 2002), major chromosome length polymorphisms were shown to be generated by meiosis (Plummer and Howlett, 1993, 1995) with significant karyotype variation linked with the assortment of parental homologues of different sizes. However, in some cases, novel-sized homologues were observed that varied in size from those of the parents (Plummer and Howlett, 1995; Leclair et al., 1996). Zolan (1995) suggested that size changes in chromosomes represented gain or loss of non-essential sequences. In *L. maculans*, large CLPs generated by meiosis were described particularly for the chromosome harbouring ribosomal DNA (rDNA) sequences (Howlett et al., 1997). Ribosomal DNA probes hybridised to chromosomal DNA ranging between 2.5 and 4.2 Mb (Figure 2), and rDNA was shown to be borne by a 1.8 Mb chromosome in the Canadian isolate Leroy (Morales et al., 1993). However, significant size polymorphism was also observed for the smallest chromosome (Leclair et al., 1996), and in the present study for other putative homologues, such as the chromosome harbouring *MinLm1*, which ranged in size from 2.15 to 2.8 Mb or the chromosome harbouring *MinLm3* which ranged in size from 1.15 to 2.5 Mb (Figure 2).

In contrast, other chromosomes only show a smaller degree of CLP, as illustrated here by the chromosome harbouring β-tubulin which ranged between 1.95 Mb and 2.3 Mb, and was previously described to range between 1.6 and 2.0 Mb in a series of Canadian isolates (Morales et al., 1993). This chromosome is characterised by the numerous genes that it is currently thought to contain and by the lack of genetic markers from the maps that could be hybridised to it (Figure 2). Leclair et al. (1996) suggested that a dispensable 'B' chromosome was present and that this was composed of repetitive sequences. Until now, no gene-coding sequences have been identified on this chromosome. Here, neither ORF-based nor repeat-based single-copy probes (e.g., minisatellites) hybridised to the minichromosome (Figure 2). Finally, mitochondrial DNA, as diffuse bands sized 100 to 150 kb, and linear plasmids (sized 9 and 10 kb) can also be resolved (Howlett, 1997).

L. maculans *genome structure as a mosaic of gene-rich and repeat-rich regions*

As yet, few regions of the genome of *L. maculans* have been sequenced and only about 300 gene sequences and Expressed Sequenced Tags (ESTs) are available on GenBank and COGEME databases (http://www.cogeme.man.ac.uk/). A 184 kb contig sequence has been obtained through the INRA-PMDV-Genoscope collaboration (Attard, 2001; Attard et al., 2002; Attard et al., 2005). Sequence data have also been obtained and analysed extensively for three gene-rich regions: 10 kb encompassing the mating type locus in two isolates (Cozijnsen and Howlett, 2003); the sirodesmin cluster region (68 kb) (Gardner et al., 2004); a 38 kb region with genes possibly involved in secondary metabolism (Idnurm et al., 2003a). This research, along with studies on particular genes or randomly mutated genes (Idnurm and Howlett, 2002; Idnurm and Howlett, 2003; Idnurm et al., 2003b; Meyer et al., 2004) has resulted in additional annotated genes published on international databases e.g. opsin, isocitrate lyase, cyanide hydratase, neutral trehalase, secreted proteins sp1 and sp2, G_α subunit (Idnurm and Howlett, 2001; Sexton and Howlett, 2000; Idnurm and Howlett, 2002; Wilson et al., 2002; Idnurm et al., 2003b; Wilson and Howlett, 2005; M. Meyer, this study) (Figure 2).

The occurrence of repeats in the *L. maculans* genome has been investigated via hybridisation of High-Density spotted BAC libraries with randomly labelled whole genomic DNA. This approach showed that 20% of the BAC clones contained high amounts of repeated DNA (Attard et al., 2005). The 184 kb contig sequence appears to be a pericentromeric region and contains the inactive Long Terminal Repeat (LTR)-retrotransposon '*Pholy*', also present in 17.1% of the genomic BAC clones (Attard et al., 2005). Comparative analysis of the two largest regions currently sequenced showed a marked difference between these two genomic regions. The sirodesmin gene cluster had 18 genes within a 55 kb region flanked by genes involved in other aspects of metabolism of *L. maculans*, very few non-coding sequences, and short intergenic regions. The sirodesmin gene cluster appears to be 'protected' from repetitive elements, although *Pholy* fragments flanked one end. In contrast to the sirodesmin gene cluster, the 184 kb region was AT-rich with many diverse and repeated elements, and only 11 open reading frames clustered within 32 kb. This mosaic of transposons was characterised by the occurrence of numerous truncated copies, including solo-LTRs, and large-scale degeneracy of the retrotransposons by Repeat Induced Point mutation (RIP) phenomenon, a mechanism by which repeated sequences are inactivated following meiosis. This process was recently shown to be active in *L. maculans* whereby copies of an introduced selectable marker (hygromycin resistance gene) were inactivated (Idnurm and Howlett, 2003). In addition to *Pholy* (*Copia*-type), at least two other *Gypsy*-type degenerate and truncated LTR-retrotransposons were clustered within this region, amounting to up to 70% of the sequence (Figure 4). Finally, preliminary data from the on-going sequencing of 1.3 Mb of the 2.8 Mb-chromosome of isolate JN3 suggest that up to 80% of the sequence consisted of large arrays of degenerate and composite repetitive elements (Gout and Ross, unpublished data; Gout, 2005).

Concluding remarks

Genetic linkage maps are still the primary tool for genome knowledge in a series of plant and animal models. For plants they also are a basic tool for positional cloning and breeding for agronomically desirable traits, including quantitative traits, eventually leading to marker-assisted selection. Due to their small genome size and laboratory tractability, fungal models such as the ascomycete yeast *Saccharomyces cerevisiae*, and the filamentous ascomycete *Neurospora crassa*, have been pioneers in the process of building genetic maps (Arnold, 1997; see also http://www.yeastgenome.org/; http://www.broad.mit.edu/annotation/fungi/neurospora_crassa_7/markers.html). This has led to maps containing hundreds of gene-based markers at a time when anonymous markers just began to emerge to be used for other models. In this respect, the development of genetic linkage maps has been very slow for other fungal species, being very often slowed down by the impossibility of achieving fertile crosses in laboratory conditions (or the lack of a sexual stage), the difficulty to visualize fungal chromosomes and the lack of genetic markers. By the beginning of the 1990s, both the development of high-throughput anonymous molecular markers and the assignment of linkage groups to chromosomes separated by pulsed field gel electrophoresis were instrumental for genetic mapping in fungi (Table 2). Genetic maps associated with physical assignment to chromosomes were developed to analyse genome structure, and to target and clone genes of interest such as avirulence/pathogenicity genes of pathogenic fungi, or quality genes for edible mushrooms, and nowadays, to prepare the numerous genome initiatives underway for numerous fungal species.

In the present study, the moderately dense genetic map presented along with preliminary assignment to chromosomes suggest that the genome of *L. maculans* consists of a mosaic of GC-equilibrated coding regions with no or few transposons, and of stretches of degenerate and truncated retrotransposons, encompassing very few ORFs. Here, we confirm that each chromosome contains stretches of degenerate retrotransposons, such as *Pholy*. We also suggest that, at least for some chromosomes, degenerate retrotransposons are clustered together, and often constitute a very large part of the chromosome whereas ORF-rich regions seem to be 'protected' from retrotransposon invasion. Apart from partial sequence data, this is substantiated by the observation that 20% of BAC clones are rich in repeats whereas other BAC clones did not hybridise to repeats, and by the fact that the genetic maps have linkage groups with clustered rep-PCR markers that are absent from larger linkage groups. In addition, the two minisatellites which resulted from sequencing of the BAC-end, chosen so that the corresponding BAC did not hybridize to repeated elements, *MinLm5* and *MinLm6* (Eckert

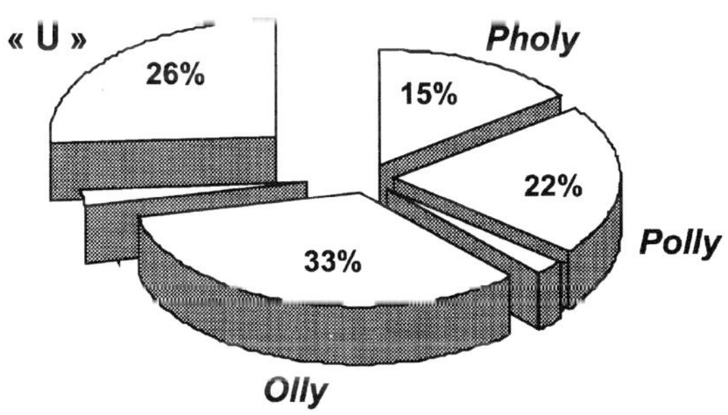

Figure 4. Sequence analysis of a 184 kb *L. maculans* genomic region (GenBank accessions AL732383, AL732384 and AL732385) expressed as the number of basepair corresponding to one given class of repeated element or 'unique' sequences (Gout, 2005). The region comprises numerous copies of more or less truncated copies of the *Copia* retrotransposon *Pholy*, and of the two recently identified *Gypsy* retrotransposons *Olly* and *Polly*. 'U' denotes non-repeated sequences either corresponding to ORFs or to other repeats of which only one copy is present in this region. Other repeats are present and correspond to a minor proportion of the sequence (minor unlabelled segments).

et al., 2005), were genetically mapped here to major LG, poor in rep-PCR markers. It remains to be seen whether large clustered arrays of repeats are a peculiarity of the *L. maculans* genome since, regardless of potential gaps in the genomic sequences, such arrays of repeats are generally not observed in the currently fully sequenced genomes of model fungi such as *S. cerivisiae* or *N. crassa*, or phytopathogenic fungi such as *M. grisea* or *Fusarium graminearum* (see www.broad.mit.edu/annotation/fungi/). The present study, along with previously published work also suggests large chromosomal length polymorphisms are linked with the loss of dispensable retrotransposon-rich regions. The dispersed nature of this repetitive DNA may play a role in generating chromosomal length polymorphisms (Plummer and Howlett, 1993, 1995) and homologues of varying sizes may be due to the presence of different numbers of repetitive sequences spaced between coding sequences, arising from unequal pairing of chromosomes. During meiosis these sequences would act as sites for reciprocal recombination, leading to production of chromosomes of novel sizes in the progeny (Plummer and Howlett, 1995). This was recently demonstrated by Gout (2005) who showed that up to 600 kb (i.e. 25% of the corresponding chromosome) consisted of mosaics of retrotransposons and other repeats, and that these regions were absent from homologues when a range of isolates was compared. This has led to the hypothesis that chromosomes may have a minimal functional size where they mainly consist of coding regions (Gout, 2005). The present study is consistent with this hypothesis and suggests that some chromosomes may be less prone to large CLP than others, suggesting they reach a minimal functional size, and are rich in ORFs. In addition, this postulate can probably be expanded to other fungal models such as *C. sativus*, where significant CLPs are noticed for the chromosome harbouring the VHv1 virulence locus (size ranging between 2.2 and 2.8 Mb) whereas other chromosomes may show only limited size polymorphism (Zhong et al., 2002).

In most fungal genome initiatives that are currently underway, a high priority is given to the construction of integrated genome maps encompassing (i) high-resolution genetic maps, (ii) physical maps, (iii) complete genomic sequence and (iv) annotation with special reference to the genes (Xu et al., 2003). Such an integrated genetic map, initiated for *L. maculans* here, is a prerequisite to understand genome structure, function and evolution of fungal genomes. The economic importance of the *L. maculans-B. napus* interaction, the properties of this interaction as a model host-pathogen system, and the need to better characterise the genome structure of the fungus has led to the establishment of the *L. maculans* genome sequence initiative. This whole-genome sequencing project is led by INRA-PMDV (T. Rouxel and M.H. Balesdent) and The University of Melbourne (B. Howlett) and will be done by the French sequencing agency Genoscope (www.cns.fr) in 2006. The objective of the project is to attain a 12-x coverage of the *L. maculans* genome, using isolate JN3 (v23.1.3) (Attard et al., 2002). The sequencing strategy will include the three following steps: (1) shot gun sequencing of chromosomal DNA of isolate v23.1.3, (2) sequencing of both ends of 12,500 BAC clones, in order to facilitate the assembly of small contigs generated by shot-gun sequences, (3) sequencing of cDNA libraries (25,000 reads), in order to facilitate and validate the automatic annotation of the sequences (Rouxel and Balesdent, 2005). Only such a large-scale sequencing project will allow us to generate the ultimate integrated maps for *L. maculans*, and enable us to validate the hypotheses on genome structure discussed in this paper based on current genetic and physical map information.

Acknowledgements

This work was partly supported by EU contracts FAIR3CT96–1669 and QLK5-CT-2002–01813 'SECURE' and also by the Australian Grains Research and Development Corporation and the Australian Research Council. Large-scale sequencing was performed in collaboration with the French sequencing agency, Genoscope (www.cns.fr). We are greatly indebted to L. Cattolico and J. Weissenbach (Genoscope, Evry, France) for the high-quality data. Thanks are due to the numerous students, post-docs and scientists who contributed to generate the INRA genetic and physical maps: F. Blaise, L. de Roquemaurel, M. Eckert, J.P. Narcy, S. Laroche, F. Parlange, S. Ross.

References

Ansan-Melayah D, Balesdent MH, Buée M and Rouxel T (1995) Genetic characterization of *AvrLm1*, the first avirulence gene of *Leptosphaeria maculans*. Phytopathology 85: 1525–1529

Arnold J (1997) Editorial. Fungal Genetics and Biology 21: 254–257

Attard A (2001) Le gène d'avirulence *AvrLm1* chez l'ascomycète phytopathogène *Leptosphaeria maculans*: Marquage moléculaire et caractérisation de son environnement génomique. Ph.D. Thesis, Institut National Agronomique, Paris, France

Attard A, Gourgues M, Gout L, Schmit J, Roux J, Narcy JP, Balesdent MH and Rouxel T (2001) Molecular characterisation and polymorphism of *MinLm1*, a minisatellite from the phytopathogenic ascomycete *Leptosphaeria maculans*. Current Genetics 40: 54–64

Attard A, Gout L, Gourgues M, Kuhn ML, Schmit J, Laroche S, Ansan-Melayah D, Billault A, Cattolico L, Balesdent MH and Rouxel T (2002) Analysis of molecular markers genetically linked to the *Leptosphaeria maculans* avirulence gene *AvrLm1* in field populations indicates a highly conserved event leading to virulence on *Rlm1* genotypes. Molecular Plant-Microbe Interactions 15: 672–682

Attard A, Gout L, Ross S, Parlange F, Cattolico L, Balesdent MH and Rouxel T (2005) Truncated and RIP-degenerated copies of the LTR retrotransposon *Pholy* are clustered in a pericentromeric region of the *Leptosphaeria maculans* genome. Fungal Genetics and Biology 42: 30–41

Aubertot JN, West J, Bousset-Vaslin L, Salam MU, Barbetti MJ and Diggle AJ (2006) Improved resistance management for durable disease control: a case study of phoma stem canker of oilseed rape (*Brassica napus*). European Journal of Plant Pathology, 114: 91–106

Balesdent MH, Attard A, Kuhn ML and Rouxel T (2002) New avirulence genes in the phytopathogenic fungus *Leptosphaeria maculans*. Phytopathology 92: 1122–1133

Berbee ML (2001) The phylogeny of plant and animal pathogens in the Ascomycota. Physiological and Molecular Plant Pathology 59: 165–187

Chalhoub BA, Thibault S, Laucou V, Rameau C, Hofte H and Cousin R (1997) Silver staining and recovery of AFLP™ amplification products on large denaturing polyacrylamide gels. BioTechniques 22: 216–220

Cozijnsen AJ and Howlett BJ (2003) Characterisation of the mating-type locus of the plant pathogenic ascomycete *Leptosphaeria maculans*. Current Genetics 43: 351–357

Cozijnsen AJ, Popa KM, Purwantara A, Rolls BD and Howlett BJ (2000) Genome analysis of the plant pathogenic ascomycete *Leptosphaeria maculansi*; mapping mating type and host specificity loci. Molecular Plant Pathology 1: 293–302

Delourme R, Chèvre A-M, Brun H, Rouxel T, Balesdent M-H, Dias JS, Salisbury PA and Rimmer RS (2006) Major gene and polygenic resistance to *Leptosphaeria maculans* in oilseed rape. European Journal of Plant Pathology, 114: 41–52

Delourme R, Pilet-Nayel ML, Archipiano M, Horvais R, Tanguy X, Rouxel T, Brun H, Renard M and Balesdent MH (2004) A cluster of major specific resistance genes to *Leptosphaeria maculans* in *Brassica napus*. Phytopathology 94: 578–583

Eckert M, Gout L, Rouxel T, Blaise F, Jedryczka M, Fitt BDL and Balesdent MH (2005) Identification and characterisation of polymorphic minisatellites in the phytopathogenic ascomycete *Leptosphaeria maculans*. Current Genetics 47: 37–48

Fitt BDL, Brun H, Barbetti MJ and Rimmer SR (2006) World-wide importance of phoma stem canker (*Leptosphaeria maculans* and *L. biglobosa*) on oilseed rape (*Brassica napus*). European Journal of Plant Patholoy, 114: 3–15

Forche A, Xu JP, Vilgalys R and Mitchell TG (2000) Development and characterization of a genetic linkage map of *Cryptococcus neoformans* var. *neoformans* using amplified fragment length polymorphisms and other markers. Fungal Genetics and Biology 31: 189–203

Gardiner DM, Cozijnsen AJ, Wilson LM, Pedras MSC and Howlett BJ (2004) The sirodesmin biosynthetic gene cluster of the plant pathogenic fungus *Leptosphaeria maculans*. Molecular Microbiology 53: 1307–1318

Gout L (2005) Clonage et évolution du gène d'avirulence *AvrLm1* chez l'ascomycète phytopathogène *Leptosphaeria maculans*. Ph.D. thesis, Institut National Agronomique, Paris-Grignon, Paris, France

Howlett BJ (1997) Genome analysis of the fungal plant pathogen *Leptosphaeria maculans* using pulsed field gel electrophoresis. Electrophoresis 18: 1544–1547

Howlett BJ (2004) Current knowledge of the interaction between *Brassica napus* and *Leptosphaeria maculans*. Canadian Journal of Plant Pathology 26: 245–252

Howlett BJ, Idnurm A and Pedras MSC (2001) *Leptosphaeria maculans*, the causal agent of blackleg disease of Brassicas. Fungal Genetics and Biology 33: 1–14

Howlett BJ, Rolls BD and Cozijnsen AJ (1997) Organisation of ribosomal DNA in the ascomycete *Leptosphaeria maculans*. Microbiological Research 152: 261–267

Idnurm A and Howlett BJ (2001) Characterization of an opsin gene from the ascomycete *Leptosphaeria maculans*. Genome 44: 167–171

Idnurm A and Howlett BJ (2002) Isocitrate lyase is essential for pathogenicity of the fungus *Leptosphaeria maculans* to canola (*Brassica napus*). Eukaryotic Cell 1: 719–724

Idnurm A and Howlett BJ (2003) Analysis of loss of pathogenicity mutants reveals that repeat-induced point mutations can occur in the Dothideomycete *Leptosphaeria maculans*. Fungal Genetics and Biology 39: 31–37

Idnurm A, Taylor JL, Pedras MSC and Howlett BJ (2003a) Small scale functional genomics of the blackleg fungus, *Leptosphaeria maculans*: analysis of a 38 kb region. Australasian Plant Pathology 32: 511–519

Idnurm A, Warnecke DC, Heinz E and Howlett BJ (2003b) Characterisation of neutral trehalase and UDP-glucose: sterol glucosyltransferase genes from the plant pathogenic fungus *Leptosphaeria maculans*. Physiological and Molecular Plant Pathology 62: 305–313

Jedryczka M, Rouxel T and Balesdent MH (1999) Rep-PCR based genomic fingerprinting of isolates of *Leptosphaeria maculans* from Poland. European Journal of Plant Pathology 105: 813–823

Jurgenson JE, Bowden RL, Zeller KA, Leslie JF, Alexander NJ and Plattner RD (2002a) A genetic map of *Gibberella zeae* (*Fusarium graminearum*). Genetics 160: 1451–1460

Jurgenson JE, Zeller KA and Leslie JF (2002b) Expanded genetic map of *Gibberella moniliformis* (*Fusarium verticillioides*). Applied and Environmental Microbiology 68: 1972–1979

Kema GHJ, Goodwin SB, Hamza S, Verstappen ECP, Cavaletto JR, Van der Lee TAJ, de Weerdt M, Bonants PJM and Waalwijk C (2002) A combined amplified fragment length polymorphism and randomly amplified polymorphism DNA genetic linkage map of *Mycosphaerella graminicola*, the septoria tritici leaf blotch pathogen of wheat. Genetics 161: 1497–1505

Lander ES, Green P, Abrahamson J, Barlow A, Daly MJ, Lincoln SE and Newburg L (1987) Mapmaker: an interactive computer package for constructing primary genetic linkage maps of experimental and natural populations. Genomics 1: 174–181

Larraya LM, Perez G, Ritter E, Pisabarro AG and Ramirez L (2000) Genetic linkage map of the edible basidiomycete *Pleurotus ostreatus*. Applied and Environmental Microbiology 66: 5290–5300

Leclair S, Ansan-Melayah D, Rouxel T and Balesdent MH (1996) Meiotic behaviour of the minichromosome in the phytopathogenic ascomycete *Leptosphaeria maculans*. Current Genetics 30: 541–548

Liew ECY, Aptroot A and Hyde KD (2000) Phylogenetic significance of the pseudoparaphyses in Loculoascomycete taxonomy. Molecular Phylogenetics and Evolution 16: 392–402

Marra RE, Huang JC, Fung E, Nielsen K, Heitman J, Vilgalys R and Mitchell TG (2004) A genetic linkage map of *Cryptococcus neoformans* variety neoformans serotype D (*Filobasidiella neoformans*). Genetics 167: 619–631

Meyer M, Blaise F, Remy E, Zhou L, Tourneur C, Narcy JP, Rouxel T and Balesdent MH (2004) *Agrobacterium tumefaciens*-mediated transformation of *Leptosphaeria maculans*. IOBC Bulletin 27: 179–189

Morales VM, Séguin Swartz G and Taylor JL (1993) Chromosome size polymorphism in *Leptosphaeria maculans*. Phytopathology 83: 503–509

Muraguchi H, Ito Y, Kamada T and Yanagi SO (2003) A linkage map of the basidiomycete *Coprinus cinereus* based on random amplified polymorphic DNAs and restriction fragment length polymorphisms. Fungal Genetics and Biology 40: 93–102

Nitta N, Farman ML and Leong SA (1997) Genome organization of *Magnaporthe grisea*: Integration of genetic maps, clustering of transposable elements and identification of genome duplications and rearrangements. Theoretical and Applied Genetics 95: 20–32

Pedersen C, Rasmussen SW and Giese H (2002) A genetic map of *Blumeria graminis* based on functional genes, avirulence genes, and molecular markers. Fungal Genetics and Biology 35: 235–246

Plummer KM and Howlett BJ (1993) Major Chromosomal Length Polymorphisms are evident after meiosis in the phytopathogenic fungus *Leptosphaeria maculans*. Current Genetics 24: 107–113

Plummer KM and Howlett BJ (1995) Inheritance of Chromosomal Length Polymorphisms in the ascomycete *Leptosphaeria maculans*. Molecular and General Genetics 247: 416–422

Pongam P, Osborn TC and Williams PH (1998) Genetic analysis and identification of Amplified Fragment Length Polymorphism markers linked to the *alm1* avirulence gene of *Leptosphaeria maculans*. Phytopathology 88: 1068–1072

Rouxel T and Balesdent MH (2005) The stem canker (blackleg) fungus, *Leptosphaeria maculans*, enters the genomic era. Molecular Plant Pathology 6: 225–241

Sexton AC and Howlett BJ (2000) Characterisation of a cyanide hydratase gene in the phytopathogenic fungus *Leptosphaeria maculans*. Molecular and General Genetics 263: 463–470

Sexton AC, Paulsen M, Woestemeyer J and Howlett BJ (2000) Cloning, characterization and chromosomal location of three genes encoding host-cell-wall-degrading enzymes in *Leptosphaeria maculans*, a fungal pathogen of *Brassica* spp. Gene 248: 89–97

Silva-Hanlin DMW and Hanlin RT (1999) Small subunit ribosomal RNA gene phylogeny of several loculoascomycetes and its taxonomic implications. Mycological Research 103: 153–160

Taylor JL, Borgmann IE and Séguin Swartz G (1991) Electrophoretic karyotyping of *Leptosphaeria maculans* differentiates highly virulent from weakly virulent isolates. Current Genetics 19: 273–277

Tzeng TH, Lyngholm LK, Ford CF and Bronson CR (1992) A Restriction Fragment Length Polymorphism map and electrophoretic karyotype of the fungal maize pathogen *Cochliobolus heterostrophus*. Genetics 130: 81–96

Vos P, Hogers R, Bleeker M, Reijans M, van der Lee T, Hornes M, Frijters A, Pot J, Peleman J, Kuiper M and Zabeau M (1995) AFLP: a new technique for DNA fingerprinting. Nucleic Acids Research 23: 4407–4414

West JS, Balesdent MH, Rouxel T, Narcy JP, Huang YJ, Roux J, Steed JM, Fitt BDL and Schmit J (2002) Colonization of winter oilseed rape tissues by A/Tox^{+} and B/Tox^{0} *Leptosphaeria maculans* (phoma stem canker) in France and England. Plant Pathology 51: 311–321

Williams RSB, Davis MA and Howlett BJ (1994) Nitrate reductase of the ascomycetous fungus, *Leptosphaeria maculans* – gene sequence and chromosomal location. Molecular and General Genetics 244: 1–8

Williams RSB, Davis MA and Howlett BJ (1995) The nitrate and nitrite reductase-encoding genes of *Leptosphaeria maculans* are closely linked and transcribed in the same direction. Gene 158: 153–154

Wilson LM and Howlett BJ (2005) *Leptosphaeria maculans*, a fungal pathogen of *Brassica napus*, secretes a subtilisin-like serine protease. European Journal of Plant Pathology 112: 112–232

Wilson LM, Idnurm A and Howlett BJ (2002) Characterization of a gene (*sp1*) encoding a secreted protein from *Leptosphaeria maculans*, the blackleg pathogen of *Brassica napus*. Molecular Plant Pathology 3: 487–493

Xu Z, Lance B, Vargas C, Arpinar B, Bhandarkar S, Kraemer E, Kochut KJ, Miller JA, Wagner JR, Weise MJ, Wunderlich JK, Stringer J, Smulian G, Cushion MT and Arnold J (2003) Mapping by sequencing the Pneumocystis genome using the Ordering DNA Sequences V3 tool. Genetics 163: 1299–1313

Zhong SB, Steffenson BJ, Martinez JP and Ciuffetti LM (2002) A molecular genetic map and electrophoretic karyotype of the plant pathogenic fungus *Cochliobolus sativus*. Molecular Plant Microbe Interactions 15: 481–492

Zolan ME (1995) Chromosome-Length Polymorphism in fungi. Microbiological Review 59: 686–698

European Journal of Plant Pathology (2006) 114: 33–40
DOI 10.1007/s10658-005-3683-5

Major gene resistance in *Brassica napus* (oilseed rape) is overcome by changes in virulence of populations of *Leptosphaeria maculans* in France and Australia

Susan J. Sprague[1,2], Marie-Hélène Balesdent[3], Hortense Brun[4], Helen L. Hayden[1], Stephen J. Marcroft[5], Xavier Pinochet[6], Thierry Rouxel[3] and Barbara J. Howlett[1,*]
[1]*School of Botany, The University of Melbourne, VIC 3010 Australia;* [2]*CSIRO Plant Industry, GPO Box 1600, Canberra ACT 2601, Australia;* [3]*INRA–PMDV Route de St Cyr, 78026 Versailles Cedex, France;* [4]*INRA Agrocampus Rennes, UMR BiO3P, Domaine de la Motte, F-35653 Le Rheu, 35327 Cedex, BP, France;* [5]*Marcroft Grains Pathology, 3400 Horsham, Victoria, Australia;* [6]*CETIOM (Centre Technique Interprofessionnel des Oléagineux Métropolitains), Scientific Direction, BP4, 78850 Thiverval-Grignon, France; *Author for correspondence (Phone: +613-8344-5062; Fax: +613-9347-5460; E-mail: bhowlett@unimelb.edu.au)*

Accepted 3 October 2005

Key words: blackleg, *Brassica rapa* ssp. *sylvestris*, *Phoma lingam*, resistance

Abstract

Resistance of *Brassica napus* (oilseed rape, canola) conferred by three different major resistance genes has been overcome by changes in virulence of *Leptosphaeria maculans* populations in France and Australia. In South Australia where *B. napus* cultivars with major gene resistance derived from *Brassica rapa* ssp. *sylvestris* were grown extensively, resistance was rendered ineffective within 3 years of commercial release of the cultivar. Disease severity was higher on cultivars with *sylvestris*-derived resistance than cultivars with polygenic resistance. This Australian situation is compared to that in France, where resistance conferred by the *Rlm1* gene was overcome nation-wide in 5 years under commercial cropping practices, and also where a source of resistance introgressed into *B. napus* from *B. juncea* was rendered inefficient in 3 years in experimental field plots near Rennes.

Introduction

Phoma stem canker (blackleg), caused by the fungus *Leptosphaeria maculans*, is the most important disease of *Brassica napus* worldwide (for reviews see Howlett, 2004; Fitt et al., 2006). As discussed by Delourme et al. (2006) and Aubertot et al. (2006), the main strategies for control of the disease are cultural practices and effective deployment of resistance genes. The durability of particular sources of resistance can vary depending on the biology of the pathogen. Pathogens that pose the greatest risk of overcoming resistance conferred by single or few genes (major gene resistance) are those with a high evolutionary potential; i.e. those that reproduce both sexually and asexually, have effective spore dispersal and large population sizes (McDonald and Linde, 2002). *Leptosphaeria maculans* has a high potential for overcoming major gene resistance since it undergoes an annual cycle of sexual recombination to produce widely dispersed wind-borne ascospores, and it also produces large numbers of conidia spread by rain splash (Fitt et al., 2006).

In accordance with this high evolutionary potential of *L. maculans*, major gene resistance has been overcome in field experiments in France (Brun et al., 2000) and in nation-wide surveys of commercial fields and trial sites (Rouxel et al., 2003). Oilseed rape breeding programs in Australia have focussed on developing cultivars displaying adult plant resistance, which provides protection only against stem canker, whilst leaves develop sporulating lesions. This non-specific resistance is probably conferred by several genes and is hereafter referred to as polygenic (for review see Delourme et al., 2006). In 2000, cultivars with specific resistance conferred by a single dominant gene were released by Pacific Seeds Pty Ltd (Li and Cowling, 2003). This resistance gene was derived from *Brassica rapa* ssp. *sylvestris* by Crouch et al. (1994) and hereafter resistance conferred by this gene is referred to as *sylvestris*-derived resistance. When these cultivars were first released, they showed few phoma leaf spots or stem canker symptoms (Sosnowski et al., 2004) and were given an Australian Blackleg Rating of 9 on a scale where 1 is highly susceptible and 9 is highly resistant (http://www.canolaaustralia.com). However, in 2002 canker symptoms were identified on such cultivars at two locations remote from each other in South-eastern Australia (Andrew Easton, Pacific Seeds Pty Ltd, Australia, pers. comm.). In addition, isolates in Western Australia capable of attacking these cultivars were identified (Li et al., 2003a; 2005).

In this paper, we report in greater detail the decline in efficiency of resistance in *B. napus* cultivars with *sylvestris*-derived resistance to *L. maculans* in South Australia and compare it to the decline in efficiency of resistance in France.

Materials and methods

Survey of commercial oilseed rape crops and trial sites in South Australia for canker severity

Following the release of cultivars with *sylvestris*-derived resistance in Australia in 2000, field surveys were conducted to detect any changes in host resistance and in *L. maculans* populations in response to the large-scale use of these cultivars. In 2003, the incidence and severity of phoma stem canker were examined in crops at different distances from a trial site on the Lower Eyre Peninsula in South Australia where diseased plants were first observed in 2002. The 2002 site (0.5 ha) had been used for the previous 3 years to assess resistance to *L. maculans* in oilseed rape lines and is in a region that had been cropped intensively with cultivars with *sylvestris*-derived resistance for the previous 3 years, in rotation with cereal or grain legume crops. Thirty-one oilseed rape crops of cultivars with *sylvestris*-derived resistance were examined within a 100 km radius of this site. These cultivars with *sylvestris*-derived resistance included Surpass 400, Surpass 501TT and Hyola 501.

In a separate experiment in 2004, disease severity in cultivars with either polygenic or major gene *sylvestris*-derived resistance was examined at 14 sites in South Australia. At one of these sites (Struan), there had been significant yield loss in cultivars with *sylvestris*-derived resistance in 2003 (Trent Potter, South Australian Research and Development Institute, Australia, pers. comm.). At most sites, the moderately resistant cv. Beacon (with polygenic resistance) and cv. Surpass 501TT were examined. Where these cultivars were not present, cultivars with a similar level of polygenic or *sylvestris*-derived resistance were assessed.

In the two experiments described above, the incidence and severity of phoma stem canker were assessed on 60 mature plants per site just before wind-rowing (swathing). Plants were cut transversely at the crown and then visually assessed for disease severity, which was scored as a percentage (0, 5, 10, 20, 30, 40, 50, 60, 70, 80, 90 or 100) of the cross section with internal necrosis (Marcroft et al., 2004). The average disease severity and incidence of plants with disease severity $\geq 5\%$ were then calculated for each field and trial site.

Virulence testing of L. maculans *isolates*

Stubble pieces bearing *L. maculans* pseudothecia from cultivars with polygenic resistance or with *sylvestris*-derived resistance were collected from Lower Eyre Peninsula and Struan in South Australia in 2002 and 2003. Individual ascospores were cultured and isolates derived from these ascospores (35 isolates from 2002 stubble and 49 isolates from 2003 stubble) were tested for virulence on the cotyledons of cv. Surpass 400. Symptoms were scored 12 or 14 days after inoculation on a scale of 0 (no darkening around wounds) to 9 (large grey-green lesions with profuse sporulation) (Koch et al., 1991). Isolates were classified as producing low (score 0 to 2), moderate (2 to 4), high (4 to 6) or very high (6 to 9) lesion scores. The percentage of isolates in each category was determined for both years.

Statistical analyses

Stem canker severity and incidence data collected at different distances from the original site where disease was observed on the Lower Eyre Peninsula in 2003, were square-root transformed and arcsin-transformed, respectively prior to analysis with REML, the algorithm for unbalanced data sets. A *t*-test was used to compare the severity of disease in cultivars with polygenic resistance to that in cultivars with *sylvestris*-derived resistance at the same site. The severity of cotyledonary lesions caused by isolates collected in 2002 and 2003 was compared using the *t*-test. All analyses were performed using Genstat version 6.1 (Payne et al., 1995).

Results and discussion

The incidence and severity of phoma stem canker in cultivars with *sylvestris*-derived resistance were examined on the Lower Eyre Peninsula in South Australia. At the sites examined there were no significant differences in the incidence and severity of disease, even at distances up to 100 km from the original 0.5 ha site where stem cankers were first observed in 2002 (Figure 1). This finding implies that individual *L. maculans* genotypes capable of overcoming *sylvestris*-derived resistance did not arise at this one location, or alternatively, that the spread of virulent isolates of the fungus was very efficient. However, it is most likely that such isolates were present over the whole area at a low frequency and were selected for by increased sowing of cultivars with *sylvestris*-derived resistance. Indeed, an isolate collected in Millicent, South Australia before cultivars with *sylvestris*-derived resistance were developed (isolate IBCN18 collected in 1986) is able to produce stem cankers on these cultivars (data not shown).

In 2004, severity of phoma stem canker was assessed on 60 mature plants at each of the 14 South Australian sites. The mean disease severity in cultivars with *sylvestris*-derived resistance was 51.7, significantly greater ($P < 0.01$) than the mean of 25.3 recorded on cultivars with polygenic resistance (Figure 2). At six of the sites, canker severity was similar in cultivars with *sylvestris*-derived and polygenic resistance, although disease severity was extremely low at two sites (Figure 2). However, at eight of the sites the pathogen populations showed host specificity whereby canker severity was higher on cultivars with *sylvestris*-derived resistance than on cultivars with polygenic resistance. This finding of populations adapted to cause cankers on cultivars with *sylvestris*-derived resistance is consistent with the extensive use of

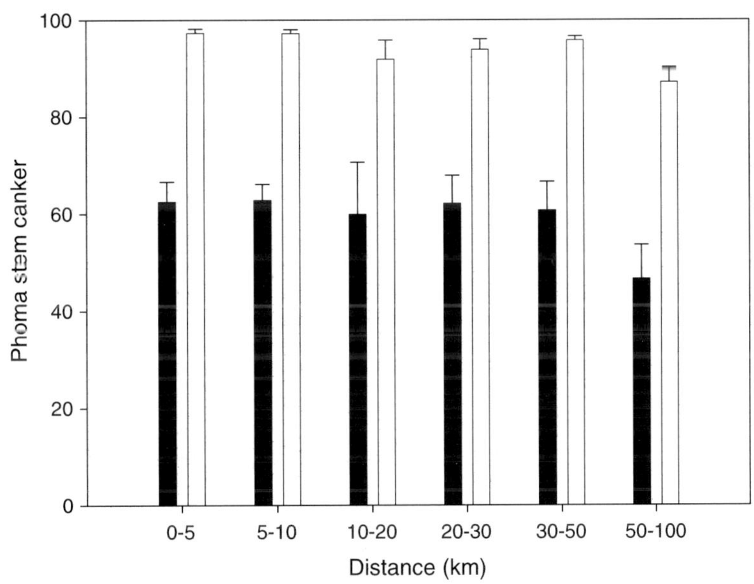

Figure 1. Severity (■) and incidence (□) of phoma stem canker in 31 commercial crops of *Brassica napus* cultivars with resistance derived from *B. rapa* subsp. *sylvestris* in 2003 at several distances from the South Australian site where plants with stem canker were first observed in 2002. Results are presented as mean ± SE.

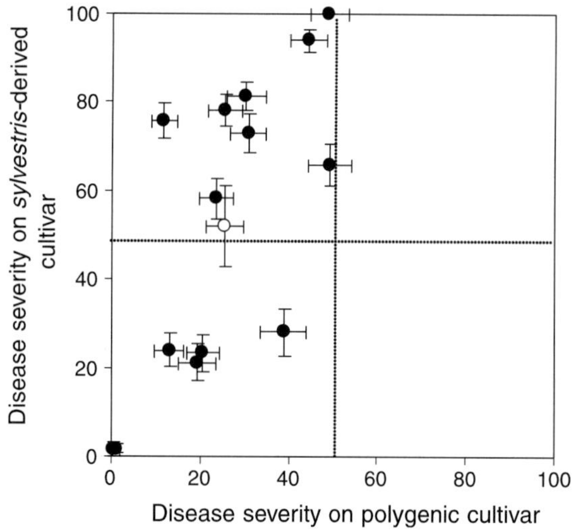

Figure 2. Severity of phoma stem canker in *Brassica napus* cultivars with either polygenic or *sylvestris*-derived resistance at 14 sites in South Australia in 2004. Each closed point (●) represents data (mean ± SE of disease severity) collected from 60 plants at one site. The open point (○) represents the mean (±SE) disease severity across all sites. Disease severity was significantly ($P < 0.01$) higher in cultivars with *sylvestris*-derived resistance than in cultivars with polygenic resistance.

such cultivars in these regions, compared to cultivars with polygenic resistance.

Individual isolates cultured from stubble of cultivars with polygenic resistance or with *sylvestris*-derived resistance were tested for their ability to cause cotyledonary lesions on cv. Surpass 400, which has *sylvestris*-derived resistance. Isolates collected in 2003 produced higher disease scores than isolates collected in 2002 (Figure 3). The mean disease score for isolates inoculated on Surpass 400 in 2003 was 4.3, significantly higher ($P < 0.001$) than the 2002 mean of 2.3 (data not shown). Although the number of isolates tested was small, this finding supports the hypothesis that the virulence of *L. maculans* isolates against *sylvestris*-derived resistance has changed. Surveys in New South Wales and Victoria have identified some fields with high levels of phoma stem canker in cultivars with *sylvestris*-derived resistance (S.J. Marcroft and S.J. Sprague, Australia, unpubl.). Isolates from Western Australia capable of overcoming the *sylvestris*-derived resistance in glasshouse trials have been described (Li et al., 2003a; 2005), but field surveys of disease severity and yield loss in this state have not yet been reported.

The situation in South Australia can be compared to the situation in France, where major gene resistance has been reported to be rendered inefficient. Brun et al. (2000) carried out a field experiment at Le Rheu over a 4-year period (1992–1995), where oilseed rape lines with major gene resistance introgressed from two different Brassica species were sown into *L. maculans*-infested stubble of the respective lines to assess the durability of the resistance sources. After 4 years of recurrent selection, the fungal populations had adapted to overcome resistance in a cultivar-specific manner. Under high inoculum concentration and in the presence of *L. maculans* populations selected recurrently on an oilseed rape line (MX) with the *Rlm6* resistance gene from *B. juncea*, resistance of this line was overcome after three seasons. In contrast, the MX-line was much more resistant to

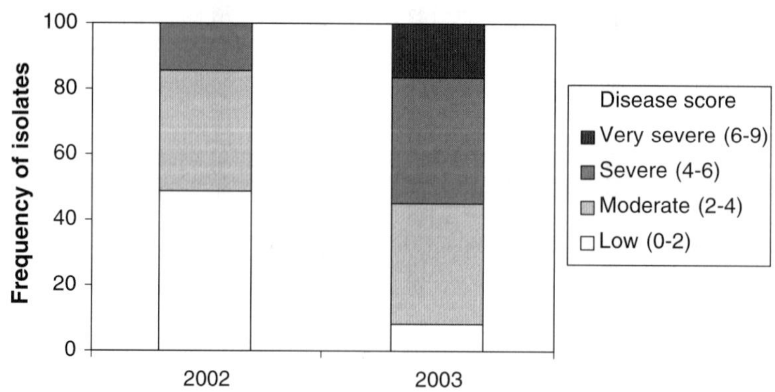

Figure 3. Frequency of *Leptosphaeria maculans* isolates collected from South Australia in 2002 and 2003 causing different levels of disease symptoms (low to very high disease scores) on the cotyledons of *Brassica napus* cv. Surpass 400. The mean disease score for isolates inoculated on Surpass 400 in 2003 was 4.3, significantly higher ($P < 0.001$) than the 2002 mean of 2.3 (data not shown).

populations selected on stubble of *B. napus* and an oilseed rape addition line (LA4+) with chromosome 4 introgressed from *B. nigra*. The resistance of the LA4+ line was maintained in the presence of *L. maculans* populations from stubble of plants with *B. juncea* or *B. nigra* resistance. Both the MX and LA4+ lines exhibited a high level of resistance when exposed to *L. maculans* populations from stubble of *B. napus* cultivars.

These recurrent selection experiments suggest that *L. maculans* populations can rapidly shift under selection pressure exerted by a new resistance gene. The *Rlm6* gene from *B. juncea* seems to have given a selective advantage to virulent isolates that pre-existed within the fungal population, as previously shown by Somda et al. (1999). Consequently, the selection pressure within the field experiment rendered the *Rlm6* resistance source inefficient after only three seasons. Unlike the *Rlm6* gene, which was introduced into a susceptible genetic background, the more stable source of resistance from *B. nigra* was introduced into the LA-line, which has polygenic resistance at the adult plant stage. Some isolates produced leaf spots but there was little canker at the stem base (data not shown). These results are consistent with the hypothesis that the durability of major resistance genes can be increased in the presence of a genetic background with polygenic resistance (Kiyosawa and Shiyomi, 1976). An alternative hypothesis is that isolates virulent on the *B. nigra* recombinant line were less fit than isolates virulent on the *B. juncea* recombinant line. As yet there are not enough data to test either of these hypotheses.

A nation-wide survey of commercial crops in France over an 8-year period (1994–2001) showed a decline in the efficiency of resistance conferred by a major gene, *Rlm1* (Rouxel et al., 2003). Increases in the frequency of virulent *avrLm1* isolates in the *L. maculans* population corresponded to the increased adoption of cultivars harbouring *Rlm1*. By 1998/1999 this led to a decrease in the effectiveness of resistance conferred by *Rlm1*. Thereafter, the area sown to cultivars harbouring *Rlm1* decreased significantly. Further large-scale surveys of the French *L. maculans* population structure in 2000/2001 confirmed that *AvrLm1* isolates represented less than 20% of the national population (Balesdent et al., 2006). Apart from the large-scale cropping of *Rlm1* cultivars, other factors probably contributed to the speed of this shift in allele frequency. These include (i) increased areas of oilseed rape in France (an increase of more than 35% between 1996 and 1999) leading to a larger pathogen population size; (ii) high inoculum concentrations in the autumns of 1998 and 1999, promoting colonisation of plants by *avrLm1* bearing-isolates and subsequent sexual recombination; and (iii) changes in cultural practices including closer crop rotations and development of minimum tillage practices. Integrated management strategies can increase the longevity of resistance (Mundt et al., 2002), however, such strategies need to be clearly communicated to growers and farm advisers (Gladders et al., 2006).

The *Rlm1*-attacking isolates were present in French *L. maculans* populations before the commercial release of cultivars with the *Rlm1* resistance gene. For example, an isolate collected at Le Rheu in 1985 and deposited in the International Blackleg of Crucifers Network (IBCN) collection (Rouxel et al., 2003) and isolate H5 obtained in 1989 (Ansan-Melayah et al., 1995) have the *avrLm1* allele. Van der Plank (1968) has proposed that the presence of a virulence gene would decrease the fitness of a pathogen, therefore the frequency of isolates harbouring that virulence gene would decline when the corresponding resistance gene was removed. This theory suggests that an isolate with an unnecessary virulence gene would be less fit and occur at very low frequency within a population. This hypothesis was tested and is discussed by Huang et al. (2006) who compared symptoms caused by near-isogenic isolates of *L. maculans* differing at the *AvrLm4* locus. Many avirulence genes are probably not associated with a fitness function and therefore a mutation to virulence is not necessarily associated with loss of fitness (Leach et al., 2001). As discussed previously, isolates able to overcome the *sylvestris*-derived resistance and *Rlm1* were present in Australia and France, respectively, prior to the introduction of cultivars harbouring the corresponding resistance gene. However, these isolates did not appear to increase in frequency until the widespread use of cultivars containing the resistance genes, indicating that virulent isolates may be less fit.

At the Lower Eyre Peninsula in South Australia, the reduced efficiency of resistance in commercial crops occurred more rapidly and was associated with greater yield loss than in France. For instance, up to 90% yield loss was reported in some fields in the Lower Eyre Peninsula but few areas in France recorded such high yield loss. Although phoma stem canker of *B. napus* is a world-wide problem, yield loss due to the disease is generally greater in Australia than in other countries. This may be because *B. napus* crops experience a milder winter with higher spring and summer temperatures in Australia compared to France. These conditions maximise the growth rate of the fungus within the plant and therefore increase the stem canker severity (Sun et al., 2001). An additional explanation for the large yield loss and rapid decline in efficiency of resistance in Australia is that the general level of background resistance of the spring oilseed rape cultivars into which the *sylvestris*-derived resistance was introduced may have been less effective than that of the European winter oilseed rape cultivars. Cultivar Surpass 400, with *sylvestris*-derived resistance has some background resistance. The severity of leaf lesions and stem canker was less on cv. Surpass 400 than on the highly susceptible cultivars Q2 and Westar when inoculated with isolates able to overcome *sylvestris*-derived resistance (Li et al., 2003b). Without specific molecular markers for resistance, the extent of polygenic resistance in a cultivar with an effective major gene resistance in field trials cannot be evaluated.

Similar changes in virulence of *L. maculans* populations leading to overcoming of resistance have not been documented in other countries. This may reflect the fact that major gene resistance in oilseed rape cultivars has only recently been described; hence any changes in field behaviour of cultivars (such as increasing disease severity) has not been attributed to a decline in efficiency of such resistance. However, major gene resistance becoming less efficient has been reported in other host–pathogen systems. *Mycosphaerella graminicola*, a foliar pathogen of wheat, and *Venturia inaequalis*, the cause of apple scab are pathogens with a similar lifecycle to *L. maculans*. These fungi undergo sexual recombination on infected debris, airborne dispersal of sexual spores and asexual multiplication within the crop, and are able to overcome major gene resistance of their hosts. Isolates of *M. graminicola* that could specifically overcome resistance of the *Stb4* major gene resistance in the wheat cv. Gene (Kronstad et al., 1994) were identified within 5 years of cultivar release in Oregon (Cowger et al., 2000). Although cv. Gene was highly resistant on release in 1992, some disease was evident (Mundt et al., 1995) suggesting that virulent isolates were already present in the population. Increased use of cv. Gene selected for virulent isolates, however, these isolates persisted in the population despite the reduction in acreage of cv. Gene (Cowger et al., 2000). In contrast to this situation with *M. graminicola*, and *L. maculans* in France and Australia, isolates of *V. inaequalis* able to overcome the *Vf* major gene resistance in apple apparently were not present in the population prior to the introduction of cultivars harbouring this resistance gene (Guerin and Le Cam, 2004). The authors propose that all virulent isolates arose from a single mutation event that occurred in the sampling year, as all virulent isolates collected could be assigned to a single clonal lineage. This hypothesis was supported by the high degree of genetic differentiation between virulent and avirulent populations, which indicated a lack of sexual recombination between the two populations (Guerin and Le Cam, 2004).

Although the oilseed rape genotype (winter vs. spring) and the environmental conditions under which crops are grown differ considerably between France and Australia, resistance conferred by major genes was overcome in a few years in both countries. The data presented for different resistance genes under diverse climatic environments highlight the need for better evaluation of the level of background resistance to *L. maculans* into which major resistance genes are introduced. Moreover, these results highlight the need to exploit the best combinations of different genetic factors (for example; by pyramiding effective major genes and/or by associating polygenic and major gene resistance) to develop cultivars with durable resistance. Delourme et al. (2006) discuss a range of resistance sources that are currently being tested for exploitation in oilseed rape breeding programs. The presence of the sexual stage of *L. maculans* on oilseed rape stubble with major resistance genes allows the recombination of virulence genes and is a source of primary inoculum for initiating epidemics. Integrated strategies that consider the genotype of the cultivars, agronomic practices to reduce fungal inoculum and the best strategy to deploy cultivars with the same major gene resistance (for example; minimum distance between fields) require further research to minimise the risk that resistance becomes inefficient in oilseed rape by changes in populations of *L. maculans*.

Acknowledgements

The research in Australia was funded by the Grains Research and Development Corporation. The research in France was funded by EU contract FAIR3CT96-1669 (IMASCORE), grants from the 'Societe interprofessionelle des oleagineux, proteagineux et cultures textiles' (SIDO), the Ágence de l 'environnement et de la maitrise de l energie' (ADEME), AIP INRA and GEVES, CETIOM, PROMOSOL and the Ministry of Agriculture.

References

Ansan-Melayah D, Balesdent MH, Buée M and Rouxel T (1995) Genetic characterisation of *AvrLm1*, the first avirulence gene of *Leptosphaeria maculans*. Phytopathology 85: 1525–1529

Aubertot JN, West JS, Bousset-Vaslin L, Salam MU, Barbetti MJ and Diggle AJ (2006) Improved resistance management for durable disease control: a case study of phoma stem canker of oilseed rape (*Brassica napus*). European Journal of Plant Pathology 114: 91–106

Balesdent MH, Louvard K, Pinochet X and Rouxel T (2006) A large-scale survey of races of *Leptosphaeria maculans* occurring on oilseed rape in France. European Journal of Plant Pathology 114: 53–65

Brun H, Levivier S, Somda I, Ruer D, Renard M and Chèvre AM (2000) A field method for evaluating the potential durability of new resistance sources: application to the *Leptosphaeria maculans–Brassica napus* pathosystem. Phytopathology 90: 961–966

Cowger C, Hoffer ME and Mundt CC (2000) Specific adaptation of *Mycosphaerella graminicola* to a resistant wheat cultivar. Plant Pathology 49: 445–451

Crouch JH, Lewis BG and Mithen RF (1994) The effect of A-genome substitution on the resistance of *Brassica napus* to infection by *Leptosphaeria maculans*. Plant Breeding 112: 265–278

Delourme R, Chèvre AM, Brun H, Rouxel T, Balesdent MH, Dias JS, Salisbury P, Renard M and Rimmer SR (2006) Major gene and polygenic resistance to *Leptosphaeria maculans* in oilseed rape (*Brassica napus*). European Journal of Plant Pathology 114: 41–52

Fitt BDL, Brun H, Barbetti MJ and Rimmer SR (2006) World-wide importance of phoma stem canker (*Leptosphaeria maculans* and *L. biglobosa*) on oilseed rape (*Brassica napus*). European Journal of Plant Pathology 114: 3–15

Gladders P, Evans N, Marcroft SJ and Pinochet X (2006) Dissemination of information about management strategies and changes in farming practices for the exploitation of resistance to *Leptosphaeria maculans* (phoma stem canker) in oilseed rape cultivars. European Journal of Plant Pathology 114: 117–126

Guerin F and Le Cam B (2004) Breakdown of the scab resistance gene *Vf* in apple leads to a founder effect in population of the fungal pathogen *Venturia inaequalis*. Phytopathology 94: 364–369

Howlett BJ (2004) Current knowledge of the *Brassica napus–Leptosphaeria maculans* interaction: a review. Canadian Journal of Plant Pathology 24: 245–252

Huang YJ, Li ZQ, Evans N, Rouxel T, Fitt BDL and Balesdent MH (2006) Fitness cost associated with the *AvrLm4* avirulence function in *Leptosphaeria maculans* (phoma stem canker of oilseed rape). European Journal of Plant Pathology 114: 77–89

Kiyosawa S and Shiyomi M (1976) Simulation of the process of breakdown of disease-resistant varieties. Japanese Journal of Breeders 26: 339–352

Koch E, Song K, Osborn TC and Williams PH (1991) Relationship between pathogenicity and phylogeny based on restriction fragment length polymorphisms in *Leptosphaeria maculans*. Molecular Plant–Microbe Interactions 4: 341–349

Kronstad WE, Kolding MF, Zwer PK and Karow RS (1994) Registration of 'Gene' wheat. Crop Science 34: 538

Leach JE, Vera Cruz CM, Bai J and Leung H (2001) Pathogen fitness penalty as a predictor of durability of disease resistance genes. Annual Review of Phytopathology 39: 187–224

Li CX and Cowling WA (2003) Identification of a single dominant allele for resistance to blackleg in *Brassica napus* 'Surpass 400'. Plant Breeding 122: 485–488

Li H, Sivasithamparam K and Barbetti MJ (2003a) Breakdown of a *Brassica rapa* subsp *sylvestris* single dominant blackleg resistance gene in *B. napus* rapeseed by *Leptosphaeria maculans* field isolates in Australia. Plant Disease 87: 752–752

Li H, Barbetti M and Sivasithamparam K (2003b) Responses of *Brassica napus* cultivars to *Leptosphaeria maculans* field isolates from Western Australia. Brassica 5: 25–34

Li H, Barbetti MJ and Sivasithamparam K (2005) Hazard from reliance on cruciferous hosts as sources of major gene-based resistance for managing blackleg (*Leptosphaeria maculans*) disease. Field Crops Research 91: 185–198

Marcroft SJ, Sprague SJ, Pymer SJ, Salisbury PA and Howlett BJ (2004) Crop isolation, not extended rotation length, reduces blackleg (*Leptosphaeria maculans*) severity of canola (*Brassica napus*) in south-eastern Australia. Australian Journal of Experimental Agriculture 44: 601–606

McDonald BA and Linde C (2002) Pathogen population genetics, evolutionary potential, and durable resistance. Annual Review of Phytopathology 40: 349–379

Mundt CC, Brophy LS and Schmitt MS (1995) Choosing crop cultivars and cultivar mixtures under low versus high disease pressure – a case study with wheat. Crop Protection 14: 509–515

Mundt CC, Cowger C and Garrett KA (2002) Relevance of integrated disease management to durable resistance. Euphytica 124: 245–252

Payne RW, Lane PW, Baird DB, Harding SA, Bicknell KE, Morgan GW, Murray DA, Thompson R, Todd AD, Tunnicliffe WG, Webster R, Welham SJ and White RP (1995) GENSTAT 5 Release 3.2 Reference Manual, Clarendon Press, Oxford, UK

Rouxel T, Penaud A, Pinochet X, Brun H, Gout L, Delourme R, Schmit J and Balesdent MH (2003) A 10-year survey of populations of *Leptosphaeria maculans* in France indicates a rapid adaptation towards the *Rlm1* resistance gene of oilseed rape. European Journal of Plant Pathology 109: 871–881

Somda I, Delourme R, Renard M and Brun H (1999) Pathogenicity of *Leptosphaeria maculans* isolates on a *Brassica napus–Brassica juncea* recombinant line. Phytopathology 89: 169–175

Sosnowski MR, Scott ES and Ramsey MD (2004) Infection of Australian canola cultivars (*Brassica napus*) by *Leptosphaeria maculans* is influenced by cultivar and environmental conditions. Australasian Plant Pathology 33: 401–411

Sun P, Fitt BDL, Steed JM, Underwood CT and West JS (2001) Factors affecting development of phoma canker (*Leptosphaeria maculans*) on stems of winter oilseed rape (*Brassica napus*) in southern England. Annals of Applied Biology 139: 227–242

Van der Plank JE (1968) Disease Resistance in Plants, Academic, London/New York

European Journal of Plant Pathology (2006) 114: 41–52
DOI 10.1007/s10658-005-2108-9

Major gene and polygenic resistance to *Leptosphaeria maculans* in oilseed rape (*Brassica napus*)

R. Delourme[1,*], A.M. Chèvre[1], H. Brun[2], T. Rouxel[3], M.H. Balesdent[3], J.S. Dias[4], P. Salisbury[5], M. Renard[1] and S.R. Rimmer[6]
[1]*INRA-Agrocampus Rennes, UMR APBV, BP 35327, 35653, Le Rheu Cedex, France;* [2]*INRA-Agrocampus Rennes, UMR BiO3P, BP 35327 & 35653, Le Rheu Cedex, France;* [3]*INRA, PMDV, Route de St-Cyr, 78026, Versailles Cedex, France;* [4]*Instituto Superior de Agronomia, Technical University of Lisbon, 1349-017, Tapada da Ajuda, Lisbon, Portugal;* [5]*Institute of Land and Food Resources, University of Melbourne, 3010, Victoria, Australia;* [6]*Agriculture and Agri-Food Canada, Saskatoon Research Centre, 107 Science Place, S7N 0X2, Saskatoon, Saskatchewan, Canada; *Author for correspondence (Phone: + 33-2-23485133; Fax: +33-2-23485120; E-mail: Regine.Delourme@rennes.inra.fr)*

Accepted 11 August 2005

Key words: blackleg, host-pathogen interaction genetics, quantitative resistance, race specific resistance genes

Abstract

The most common and effective way to control phoma stem canker (blackleg) caused by *Leptosphaeria maculans* in oilseed rape (*Brassica napus*) is through the breeding of resistant cultivars. Race specific major genes that mediate resistance from the seedling stage have been identified in *B. napus* or have been introgressed from related species. Many race specific major genes have been described and some of them are probably identical in *B. napus* (allotetraploid AACC) and the parental species *B. rapa* (diploid AA). More work is needed using a set of well-characterised isolates to determine the number of different major resistance genes available. In some *B. napus* cultivars, there is resistance which is polygenic (mediated by Quantitative Trait Loci) and postulated to be race non-specific. Many of these major genes and Quantitative Trait Loci for resistance to *L. maculans* have been located on *B. napus* genetic maps. Genes involved in race specific and polygenic resistance are generally distinct.

Abbreviations: LG – Linkage Group; QTL – Quantitative Trait Loci; RGA – Resistance Gene Analogues

Introduction

This review focuses on resistance of oilseed rape (*Brassica napus* L.) to *Leptosphaeria maculans* Desm. (Ces & de Not) since this pathogen causes more damage to oilseed rape than that caused by other members of the *Leptosphaeria* species complex, such as *L. biglobosa* (Shoemaker and Brun, 2001), that are found on Brassicaceae species. Furthermore, only the genetics of the interaction between *L. maculans* and *Brassica* spp. has so far been studied.

Different sources of resistance to *L. maculans* have been identified and introduced into *B. napus* breeding lines and cultivars. Many studies on the inheritance of resistance have been done at both seedling and adult plant growth stages. Two types of resistance are usually distinguished. The first type is a qualitative resistance, which is expressed from the seedling to the adult plant stage in cotyledons and leaves and is generally considered as single-gene race specific resistance. The second type is a quantitative adult-plant resistance, which is a partial resistance usually thought to be race non-specific and mediated by many genes. In Europe, Canada and Australia, many resistant cultivars have been registered but there is evidence of breakdown of race specific resistance in response to rapid evolution of *L. maculans* populations. Therefore, understanding the genetic basis of resistance in oilseed rape is strategically important for management of resistant cultivars. This paper reviews knowledge on race specific and race non-specific resistance and the relationship between them.

Race specific resistance genes in different Brassica *species*

Differential interactions in the *Brassica – L. maculans* pathosystem were first studied at the seedling stage using a cotyledon inoculation test (Williams and Delwiche, 1979). The first *B. napus* differential set consisted of three cultivars, 'Westar' (susceptible control, spring oilseed type), 'Quinta' and 'Glacier' (winter oilseed types) (Mengistu et al., 1991). Using this differential set, *L. maculans* isolates were classified into three Pathogenicity Groups (PG), i.e. PG2 (avirulent on 'Quinta' and 'Glacier'), PG3 (avirulent on 'Quinta' but virulent on 'Glacier') and PG4 (virulent on all three cultivars). Badawy et al. (1991) replaced 'Westar' with winter *B. napus* cultivar 'Lirabon' and added 'Jet Neuf', leading to the description of six PG, termed A1-A6, resulting from a subdivision of each of the previous groups into two PG (virulent or avirulent on 'Jet Neuf', respectively). Other race specific interactions were described using other differential sets including other *Brassica* species (Cargeeg and Thurling 1980; Ballinger et al., 1991; Kutcher et al., 1993; Kuswinanti et al., 1999). Genetic studies demonstrated a number of gene-for-gene interactions between *B. napus* and *L. maculans* and both avirulence genes (*AvrLm*) in the pathogen and their corresponding resistance genes (*Rlm*) in the host have been identified. Race specific resistance to isolates of *L. maculans* with the corresponding avirulence allele results in an incompatible interaction that inhibits infection from germinated ascospores or conidia and subsequent development of leaf lesions.

Genes identified in B. napus

The first race specific resistance genes were identified in 'Quinta' and 'Glacier' cultivars in the original differential set (Rimmer and van den Berg, 1992). Gene-for-gene *B. napus*/*L. maculans* interactions (*Rlm1*/*AvrLm1* in 'Quinta'-PG3; *Rlm2*/*AvrLm2* in 'Glacier'-PG2 interactions) were demonstrated through the use of segregating populations of both plant and pathogen (Ansan-Melayah et al., 1995, 1998). Other dominant race specific resistance genes have been described through genetic studies involving different oilseed rape cultivars/lines and different *L. maculans* isolates (Table 1). Some of these genes have been positioned on *B. napus* linkage maps (Ferreira et al., 1995; Mayerhofer et al., 1997; Delourme et al., 2004; Rimmer, 2006).

Mapping studies showed that some of the resistance genes are organised in clusters. Zhu and Rimmer (2003) found two closely linked but distinct loci mediating resistance at the seedling and adult plant stage, respectively in two *B. napus* breeding lines ('RB87-62' and 'DH88-752'). These genes all mapped to Linkage Group 6 (LG6) of the genetic map published by Ferreira et al. (1994). The two resistance loci in line 'RB87-62' mapped more than 40 cM away from those in line 'DH88-752', but only 5-10 cM separated the seedling and adult plant resistance loci of each line (Rimmer, 2006). Three other resistance genes (*LEM1*, *LmR1* and *cRLMm*, present in 'Major', 'Shiralee' and 'Maluka', respectively) have also been mapped onto this linkage group (Ferreira et al., 1995; Mayerhofer et al., 1997; Rimmer, 2006). From comparing their locations on the LG6 linkage group, it seems that *LmR1* is different from *LEM1* but *LmR1* could be identical to *cRLMm* since 'Shiralee' and 'Maluka' share a similar pedigree and produce similar interactions with *L. maculans* (Mayerhofer et al., 1997). Based on differential interactions with a series of *L. maculans* isolates, it seems likely that the seedling resistance genes in 'Maluka' (*cRLMm*) and 'RB87-62' (*cRLMrb*) are equivalent (Rimmer, 2006).

Delourme et al. (2004) have mapped five race specific resistance genes (*Rlm1, Rlm3, Rlm4, Rlm7* and *Rlm9*) on LG10 and one gene (*Rlm2*) on LG16 of the genetic map published by Lombard and Delourme (2001). *Rlm1* is clearly distinct from *Rlm3* and *Rlm4* because they both occur in one cultivar and they map to different positions. *Rlm3* and *Rlm4* are found in many cultivars but rarely seem to be present together in a single cultivar. Similarly, *Rlm3* and *Rlm7* have not been

Table 1. Genetic interactions between *Brassica napus* (allotetraploid AACC) and *Leptosphaeria maculans* including race specific resistance genes identified in *B. napus*

Cultivar/line	*L. maculans* isolate		Resistance gene	Location in *B. napus* LG[c]	Reference
	Name & information[a]	Genotype[b]			
Quinta	11.26.11 (PG3; A2)	***AvrLm1*** *avrLm2 avrLm3 avrLm4 AvrLm7 avrLm9*	*Rlm1*	LG10=N7	Ansan-Melayah et al. (1998) Delourme et al. (2004)
Maxol	v11.1.2 (PG3; A2)	***AvrLm1*** *avrLm2 avrLm3 avrLm4 AvrLm7 avrLm9*	*Rlm1*	LG10=N7	Balesdent et al. (2002) Delourme et al. (2004)
Glacier	14.3.01 (PG2; A4)	*AvrLm1* ***AvrLm2*** *avrLm3 avrLm4 AvrLm7 avrLm9*	*Rlm2*	LG16=N10	Ansan-Melayah et al. (1998) Delourme et al. (2004)
Major	PHW1245 (PG2; A3)	*AvrLm1 AvrLm2 avrLm3* ***AvrLm4*** *AvrLm7 avrLm9*	*LEM1* = *Rlm4*	LG6=N7	Ferreira et al. (1995)
Shiralee	Canadian isolates	–	*LmR1* = *Rlm4 ?*	LG6=N7	Mayerhofer et al. (1997)
Crésor	Field population (Saskatchewan)	–	$LmFr_1$	–	Dion et al. (1995)
Crésor	Field population (Saskatchewan)	–	*aRLMc*	LG6=N7	Rimmer (2006)
Maluka	Pl86.12 (PG2; –)	–	*CRLMm* = *Rlm4?*	LG6=N7	Rimmer (2006)
RB87–62	Pl86.12 (PG2; –)	–	*CRLMrb* = *Rlm4? aRLMrb*	LG6=N7	Rimmer (2006) Zhu and Rimmer (2003)
DH88–752	Pl86.12 (PG2; –)	–	*cRLMj aRLMj*	LG6=N7	Rimmer (2006) Zhu and Rimmer (2003)
Quinta	v23.2.1 (PG4; A5)	*avrLm1 avrLm2 avrLm3* ***AvrLm4*** *AvrLm7 avrLm9*	*Rlm4*	LG10=N7	Balesdent et al. (2001) Delourme et al. (2004)
Maxol	19.2.01 (PG4; A1)	*avrLm1 avrLm2* ***AvrLm3*** *avrLm4 avrLm7 avrLm9*	*Rlm3*	LG10=N7	Balesdent et al. (2002) Delourme et al. (2004)
23.1.1	A290 (PG4; A1)	*avrLm1 avrLm2 avrLm3 avrLm4* ***AvrLm7*** *avrLm9*	*Rlm7*	LG10=N7	Balesdent et al. (2002) Delourme et al. (2004)
Darmor	IBCN56	*AvrLm1? AvrLm2? AvrLm3? avrLm4 avrLm7* ***AvrLm9***	*Rlm9*	LG10=N7	Balesdent et al. (2002) Delourme et al. (2004)

[a]Pathogenicity groups are indicated as PG2-PG4 (Mengistu et al., 1991) and as A1-A6 (Badawy et al., 1991).
[b]*AvrLm* genes matching the *Rlm* genes studied are indicated in bold typeface.
[c]Linkage groups LG10 and LG16 are from the Lombard and Delourme (2001) genetic map; LG6 is from the Ferreira et al. (1994) genetic map; LG N7 and N10 are from the Parkin et al. (1995) genetic map.

found in the same cultivar. Thus, *Rlm3, Rlm4, Rlm7* and *Rlm9* could be a cluster of tightly linked genes, or a single gene with different alleles, or a combination of both. Both LG6 of the genetic map published by Ferreira et al. (1994) and LG10 published by Lombard and Delourme (2001) seem to correspond to LG N7 of the genetic map described by Parkin et al. (1995). Thus, some of the genes described on these LG might be the same. The genes *LEM1* and *cRLMm* are almost certainly identical to *Rlm4*, present in 'Major' and 'Maluka' (Rouxel et al., 2003). Additionally, *LmR1* in 'Shiralee' and *cRLMrb* in 'RB87-62' might also correspond to *Rlm4*. The different locations of *LmR1* and *LEM1* (Mayerhofer et al., 1997) might be due to the homeologous reciprocal translocation that can occur between LG N16 and LG N7 close to the position of *LEM1* (Osborn et al., 2003). Such homeologous reciprocal translocation can affect recombination and precise mapping in this region using parents with or without the translocation.

Definite conclusions on identity of or distinctness between these *Rlm* genes will be possible only through a precise characterisation of *B. napus*/*L. maculans* interactions using differential *L. maculans* isolates selected or genetically bred to carry single (or as few as possible) identified avirulence (*Avr*) genes (Balesdent et al., 2002), through allelism tests or, in the longer term, by cloning and sequence comparison of the resistance genes. An improved host differential set comprising fixed cultivars or lines possessing a minimum number of *Rlm* genes has been developed (Balesdent et al., 2005). It consists of 'Westar' (no R genes, susceptible control), 'Columbus' (*Rlm1-Rlm3*), 'Bristol' (*Rlm2-Rlm9*), '22-1-1' (*Rlm3*), 'Jet Neuf' (*Rlm4*), '150-2-1' (*B. juncea* line, *Rlm5*, not yet characterised at the *Rlm9* locus), 'Darmor-MX' (*Rlm6*, not yet characterised at the *Rlm9* locus), '23-1-1' (*Rlm7*), '156-2-1' (*B. rapa* line, *Rlm8*, not yet characterised at the *Rlm9* locus) and 'Goeland' (*Rlm9*). The host genotypes carrying genes originating from *B. napus* are freely available to the scientific community, so that a common nomenclature can be used to simplify the identification of genes for resistance to *L. maculans* in different genotypes.

Genes identified in other Brassica *species*

Resistance to *L. maculans* in germplasm of other Brassicaceae species related to *B. napus* has also been studied. Few resistance genes were found by screening different accessions of the two diploid progenitors of oilseed rape, *B. oleracea* (CC, $2n=18$) and *B. rapa* (AA, $2n=20$). An extensive screening of *B. oleracea* germplasm in the main European Gene Banks was done at 'Instituto Superior de Agronomia' (ISA Lisbon). The differential isolates were BBA62908, harbouring *AvrLm1*, *AvrLm2* and *AvrLm4* alleles (Rouxel et al., 2003), and three 'PG4' European isolates harbouring none of these avirulence alleles. Of the 392 accessions tested, a few occasionally reacted to one of the 'PG4' isolates, but none was resistant to the isolate BBA62908, suggesting the absence of *Rlm1*, *Rlm2* or *Rlm4* in *B. oleracea* genotypes (JS Dias, unpubl.). These data, which are consistent with the data of Mithen et al. (1987) and Rimmer and van den Berg (1992), confirm that no major resistance genes to *L. maculans* originate from *B. oleracea*. However, in one closely related species, *B. insularis* ($2n=18$), two dominant resistance genes were detected in a segregating population obtained from a *B. oleracea* × *B. insularis* hybrid (Mithen and Lewis, 1988).

This screening of genetic resources also encompassed 555 accessions of *B. rapa*, including accessions of vars *chinensis*, *japonica*, *parachinensis*, *pekinensis*, *perviridis*, *rapifera* and *trilocularis* and a few wild accessions. Most (95.5%) of these accessions were fully susceptible to all four *L. maculans* isolates. However, 12 (2%) accessions were resistant to all four isolates and 10 (1.8%) accessions were resistant to isolate BBA62908 and susceptible to the three 'PG4' isolates (JS Dias, unpubl.). These data suggest that the resistant accessions of *B. rapa* could harbour genes previously identified in *B. napus* such as *Rlm1*, *Rlm2* or *Rlm4*. To test this hypothesis, limited screening was done through collaboration between IPK Gatersleben and INRA-PMDV (T Rouxel and E Willner, unpubl.). Sixty-two *B. rapa* var. *oleifera* accessions, a few *B. rapa* var. *sylvestris* accessions and wild accessions were inoculated with differential isolates BBA62908 [race Av1-2-4-5-6-7-(8)], v11.1.1 [Av5-6-7-8], v11.1.2 [Av1-5-6-7-8] and v23.2.1 [Av4-5-6-7-8]. Twenty-two percent of the accessions were susceptible to all isolates and 48.3% of the accessions showed either a heterogeneous or a homogeneous resistance to all four isolates (T Rouxel and MH Balesdent, unpubl.). Of these, four accessions have *Rlm1*, three accessions have *Rlm4* and two accessions have both genes. The resistant accessions were investigated using a wide range of differential isolates. In at least one accession (CR1478), self pollination of one fully resistant plant generated a line expressing the *Rlm7* resistance. Screening of progeny of another resistant accession (156.1.1) showed monogenic control by *Rlm8* interacting with the novel single-gene avirulence *AvrLm8* (Balesdent et al., 2002). In a few accessions, resistance was observed against all or most isolates tested, suggesting occurrence of undescribed major resistance genes.

Dominant resistance genes were also identified in two *B. rapa* cultivars (Crouch et al., 1994; Chèvre et al., 2003) and a cluster of race specific genes, effective at the cotyledon stage, was identified in one source (Chèvre et al., 2003). These *B. rapa* genes were introduced into the *B. napus* genome either through production of a synthetic oilseed rape crossed to *B. napus* (Crouch et al., 1994) or by direct crosses between *B. napus* and *B. rapa* (Chèvre et al., 2003) (Table 2). The efficiency of the different introgression methods is under study (AM Chèvre, unpubl.). Genetic studies with lines obtained from the synthetic *B. napus* indicated the presence of three genes introgressed from *B. rapa* var. *sylvestris* on different *B. napus* linkage groups; *LepR1* and *LepR2* were mapped, respectively, onto *B. napus* A-genome LG N2 and LG N10 of the Parkin et al. (1995) genetic map (Yu et al., 2005), and *LepR3* was identified from new commercial cultivars Surpass 400 (Li and Cowling, 2003) and Hyola 60. *LepR3* was mapped onto *B. napus* LG N10 about 15 cM below *LepR2* (Yu et al., 2004). The LG N10 is the LG where *Rlm2* mapped (Delourme et al., 2004). In seedling assays, *LepR1* behaved as a dominant allele and was resistant to all except one *L. maculans* isolates. The *LepR3* gene was described as a dominant gene (Li and Cowling, 2003), whereas *LepR2* was incompletely dominant to most isolates, with the phenotype of the heterozygotes more similar to that of the susceptible parent than to that of the homozygous resistant lines. Isolates virulent on *LepR2* have been identified (F Yu, SR Rimmer and DJ Lydiate, unpubl.). Resistance conferred by *LepR3* has been overcome in some parts of Australia (Li et al., 2003; Sprague et al., 2006). Thus these three genes are race specific. A recessive gene has also been identified in *B. napus* lines derived from *B. rapa* var. *sylvestris*. Mapping of this locus is in progress (S R Rimmer, unpubl.). The cluster of race specific dominant *B. rapa* resistance genes (Chèvre et al., 2003) has been transferred into *B. napus* genetic backgrounds with or without polygenic resistance and is being tested under field conditions. This cluster was introgressed into a different *B. napus* linkage group (AM Chèvre, unpubl.).

The *Brassica* species with the B genome, *B. nigra* (BB, $2n=16$), *B. juncea* (AABB, $2n=36$) and *B. carinata* (BBCC, $2n=34$) have been described as highly resistant to *L. maculans* under field conditions (Rimmer and van den Berg, 1992). Based on cotyledon and stem resistance ratings, Keri et al. (1997) suggested that resistance in *B. juncea* is mediated by two genes. This is consistent with genetic data obtained with *L. maculans*, which showed that the interaction was governed by two avirulence genes termed *AvrLm5* and *AvrLm6* (Balesdent et al., 2002). The corresponding resistance genes were fixed, respectively, in a *B. juncea* line originating from 'Aurea' (*Rlm5*) and in the series of introgressed *B. napus* MX lines developed at INRA Rennes (*Rlm6*) (Chèvre et al., 1997; Balesdent et al., 2002, 2005). The resistance genes were introgressed into *B. napus* either by hand pollination between the donor species and *B. napus* cultivars/lines or by symmetric or asymmetric protoplast fusion (Table 2) and the resulting hybrids were backcrossed to *B. napus*. Whatever the screening methods used, all genes detected were dominant, except for one recessive gene introgressed from *B. juncea* (Saal et al., 2004) and three genes acting in a complex interaction (Pang and Halloran, 1996a). Evaluation of different *B. napus*-*B. nigra* addition lines carrying resistance has suggested that a number of different resistance genes occur in the B genome (Zhu et al., 1993; Chèvre et al., 1996). Resistance genes, introgressed from *B. nigra, B. juncea* or *B. carinata* into the *B. napus* genome are all on the same B genome region (Dixelius, 1999). Furthermore, Plieske et al. (1998) found that resistance genes from these three species all introgressed into the same *B. napus* linkage group. In all the introgression lines obtained by sexual crosses, resistance genes from the B genome were introgressed into A genome linkage groups of *B. napus* (Roy, 1978; Barret et al., 1998; Plieske et al., 1998). However, from their location on *B. napus* genetic maps, it seems that different genes were introgressed. This result was confirmed by the different interactions from different introgressed lines (Saal et al., 2004; AM Chèvre, unpubl.).

Other sources of resistance are available in less closely related species such as *Arabidopsis thaliana, Sinapis arvensis, Coincya monensis, Diplotaxis muralis, Diplotaxis tenuifolia* or *Raphanus raphanistrum* (Chen and Seguin-Swartz, 1999; Winter et al., 1999; Snowdon et al., 2000; Bohman et al., 2002). Some gene introgressions have been attempted by crosses to *B. napus* or by asymmetric protoplast fusion for *Arabidopsis* (Table 2). Resistant addition lines have been obtained from *B. napus*–*S. arvensis* hybrids (Snowdon et al., 2000). Bohman et al. (2002) showed that an introgression of genes carried by chromosome 3 of *A. thaliana* confers adult leaf resistance in *B. napus*.

Table 2. Introgression of resistance genes from related species into a *Brassica napus* genetic background

Donor species		Resistance tests	Genetic control	References
Diploid species	*B. rapa* (AA, $2n=20$)	Cotyledon, leaf, field	Dominant gene	Crouch et al. (1994)
		Field	Dominant gene	Li and Cowling (2003)
		Cotyledon	Dominant genes	Chèvre et al. (2003)
		Cotyledon, field	Two genes, *LepR1* and *LepR2*	Yu et al. (2005)
	B. nigra (BB, $2n=16$)	Leaf, stem	–	Sjödin and Glimelius (1989)*
		Petiole	Three additional chromosomes	Zhu et al. (1993)
		Cotyledon, field	One additional chromosome	Chèvre et al. (1996)
		Petiole	Dominant gene, *PhR1*	Plieske et al. (1998)
		Leaf	Two independent dominant genes, *LmBR2* and *LmBR3*	Dixelius (1999)*
		Cotyledon, leaf	Three independent genes	Dixelius and Whalberg (1999)*
		Stem, field	–	Ogbonnaya et al. (2003)
	Sinapis arvensis (SarSar, $2n=18$)	Cotyledon, stem	One additional chromosome	Snowdon et al. (2000)
	Arabidopsis thaliana (AtAt, $2n=10$)	Leaf	–	Bohman et al. (2002)
Allotetraploid species	*B. juncea* (AABB, $2n=36$)	Cotyledon, field	Dominant gene(s)	Roy (1978, 1984)
		Seed, cotyledon, leaf	Dominant character	Sacristan and Gerdemann (1986)
		Leaf, stem	–	Sjödin and Glimelius (1989)*
		Stem	Three genes with interaction	Pang and Halloran (1996a)
		Cotyledon, field	Dominant gene, *JLm1*	Chèvre et al. (1997), Barret et al. (1998)
		Petiole	Dominant gene, *PhR2*	Plieske et al. (1998)
		Cotyledon, stem	–	Winter et al. (1999)
		Leaf	Dominant gene, *LmBR1*	Dixelius (1999)*
		Cotyledon, leaf	Three independent genes	Dixelius and Whalberg (1999)*
		Cotyledon	Recessive gene, r_jlm2	Saal et al. (2004)
	B. carinata (BBCC, $2n=34$)	Leaf, stem	–	Sjödin and Glimelius (1989)*
		Petiole	Dominant gene, *PhR3*	Plieske et al. (1998)
		Cotyledon, leaf	Three independent genes	Dixelius and Whalberg (1999)*

*Material produced from symmetric and asymmetric protoplast fusions; – no information.

Correlation between race specific resistance at seedling and adult stages

Comparisons between seedling (cotyledon test) and adult (petiole or stem inoculation in glasshouse or field tests) resistance screening tests have produced either significant positive (McNabb et al., 1993; Bansal et al., 1994) or non-significant (Ballinger and Salisbury, 1996; Pang and Halloran, 1996b) correlations. These differences may be explained by differences between sources of resistance studied (conferring either race non-specific quantitative resistance *versus* race specific resistance or a combination of both resistance types) and differences in combinations of avirulence genes between *L. maculans* isolates used in controlled environment tests and *L. maculans* populations in field tests. Another explanation is that isolates may interact with each other. For example, Mahuku et al. (1996) reported that the weakly virulent *L. biglobosa* can induce resistance in *B. napus* to the highly virulent *L. maculans*.

The effectiveness of race specific resistance genes at growth stages later than seedlings has been clearly demonstrated. The effect of *LEM1* in 'Major' was detected using a stem inoculation test (Ferreira et al., 1995). The *LmFr1* gene from 'Crésor' accounted for 57–84% of the variation in resistance in a segregating doubled haploid (DH) population in field trials, depending on the year/location of the trial (Dion et al., 1995). The *Rlm1* gene in 'Maxol' explained 70% of the phenotypic variation for resistance in a field trial (Delourme et al., 2004). Currently, the *Rlm7* resistance is 100% effective in France because nearly 100% of field isolates of *L. maculans* harbour *AvrLm7* (Balesdent et al., 2005). Similarly, cultivars/lines with race specific resistance genes introgressed from *B. rapa* var. *sylvestris* (*LepR1*, *LepR2* and *LepR3*) were highly resistant to *L. maculans* in field trials (Li and Cowling, 2003; Yu et al., 2005), except for those with *LepR3*, which has been overcome, so that large yield losses have occurred in regions of south eastern Australia (Sprague et al., 2005). Furthermore, resistance genes introgressed into *B. napus* from *B. nigra* (Chèvre et al., 1996) or *B. juncea* (Roy, 1984; Chèvre et al., 1997) are generally effective in field trials.

Conversely, Zhu and Rimmer (2003), comparing the results of cotyledon and stem inoculation tests on lines 'RB87-62' and 'DH88-762', concluded that distinct but linked genes were effective in each line at the seedling and adult stages. The effect of *LEM1* was not detected in field trials where *L. maculans* isolates were predominantly of the same pathogenicity group (PG2) as the isolate used to identify this gene at the seedling stage. However, some isolates in the *L. maculans* field population were highly virulent on 'Major'. A difference in avirulence allele composition between *L. maculans* isolates could explain the contrasting response of 'Major' in controlled environment and field experiments (Ferreira et al., 1995) since PG2 isolates can be either virulent or avirulent on lines with resistance conferred by *Rlm4* (Badawy et al., 1991). Consequently, in the field, the effect of a race specific resistance gene will depend on the *L. maculans* population structure, i.e. on the frequency of the corresponding avirulence allele. However, the threshold frequency of the virulence allele at which the corresponding resistance gene is no longer effective in protecting the crop is not known (Brun et al., 2004).

Quantitative resistance in B. napus

A high level of field resistance to *L. maculans* in the absence of effective race specific resistance genes has been observed in winter European *B. napus* cultivars such as 'Jet Neuf', which is one of the best known sources of quantitative resistance to *L. maculans*. Cultivar Jet Neuf was widely grown all over Europe during the 1970s and 1980s and is still very resistant to *L. maculans*. The major sources of resistance used in the Australian *B. napus* breeding programmes have been Japanese spring types and French winter types (Roy et al., 1983). Although Japanese lines such as 'Chikuzen' and 'Chisaya' are only moderately resistant to *L. maculans*, resistant selections from crosses between these and other lines were obtained. Two other Japanese cultivars ('Norin 20' and 'Mutu') also showed resistance and have been widely used in breeding programmes (Salisbury and Wratten, 1999). There is usually no difference in the development of phoma leaf spot symptoms on young plants between cultivars with quantitative resistance to *L. maculans* and cultivars without it, but later in the season stem cankers do not develop or are less severe on the cultivars with quantitative resistance than those without it. *L. maculans* can survive and reproduce on even the most resistant lines (Marcroft et al., 2004). As quantitative resistance is partial, when *L. maculans* inoculum concentrations are high, it may not prevent large yield losses (Salisbury et al., 1995; Khangura and Barbetti, 2001; Marcroft et al., 2003).

Screening for quantitative resistance is primarily done by assessment of stem cankers on mature plants in field nurseries where plants have been exposed to the locally prevalent mixture of *L. maculans* races. Phoma stem canker severity is assessed using a disease index based on the extent of external and internal necrosis at the crown (stem base) of plants sampled just before harvest. Controlled environment tests for quantitative resistance using inoculation of leaves, petioles or stems with *L. maculans* have also been proposed (Newman and Bailey, 1987; Kutcher et al., 1993; McNabb et al., 1993; Bansal et al., 1994; Ballinger and Salisbury, 1996; Pang and Halloran, 1996b). With these tests, the correct evaluation of the quantitative resistance of a *B. napus* genotype depends on the *L. maculans* isolate used. Since the effect of a race specific resistance gene is detectable at later growth stages, *L. maculans* isolates that are virulent against any race specific resistance gene(s) present in the genotype to be tested must be used. Similarly, a cultivar carrying a new race specific resistance gene that is effective against all or most of the *L. maculans* isolates in a field population cannot be evaluated for quantitative resistance in that field. A controlled environment test can be done, provided a *L. maculans* isolate virulent against that particular race specific resistance gene is used.

Little information is available on the genetic control of quantitative resistance to *L. maculans*. Ferreira et al. (1995) detected two QTL, which were associated with field resistance in Manitoba, on LG12 and LG21. The genetic basis of quantitative resistance in the French winter oilseed rape 'Darmor', derived from 'Jet Neuf', has been studied. In the 'Darmor-*bzh*'×'Yudal' cross, Pilet et al. (1998) identified a total of ten QTL for resistance, of which four were associated with decreased stem canker severity and decreased plant death in both seasons of field experiments. Analysis of progeny derived from a 'Darmor'×'Samourai' cross, consisting of one DH population and a number of $F_{2:3}$ families, identified six QTL in the DH population and four QTL in the $F_{2:3}$ families (Pilet et al., 2001). Out of a total of sixteen loci detected in the four cultivars, only four QTL were common to the 'Darmor-*bzh*'×'Yudal' and 'Darmor'×'Samourai' crosses. Pilet et al. (2001) concluded that the genetic background contributes greatly to the observed QTL and that the concentration of *L. maculans* inoculum at each location is probably important in revealing QTL with small contributions to overall field resistance to *L. maculans*.

The genomic regions carrying the most consistent resistance QTL in 'Darmor' do not correspond to the two regions on LG10 and LG16 identified as carrying race specific resistance genes to *L. maculans* (Delourme et al., 2004). The position of *Rlm2* on LG16 corresponds to a QTL identified for adult plant resistance in the 'Darmor'×'Samourai' DH population (Pilet et al., 2001). The cultivar Samourai carries both the resistance allele at this QTL and *Rlm2*. Since no French isolates of *L. maculans* carry *AvrLm2* (Rouxel et al., 2003), two hypotheses can be proposed to explain this co-location; either the *Rlm2* gene has a residual effect at the adult plant stage, similar to that suggested in other pathosystems, or genes linked to *Rlm2* are responsible for part of the variation for resistance at this QTL.

Towards identification of the function of resistance genes

Although resistance genes have been cloned from many plant species, including the model species *A. thaliana*, none has yet been characterised in *Brassica* species. Expressed Sequence Tags (ESTs) were derived from *B. napus* 'Glacier' leaves inoculated with a *L. maculans* PG2 isolate (Fristensky et al., 1999). Resistance gene analogues (RGA), either derived from ESTs that have sequence homology to cloned resistance genes or from PCR products amplified with primers based on the conserved nucleotide binding site and leucine-rich repeat regions of cloned genes, have been mapped in *B. napus* (Joyeux et al., 1999; Sillito et al., 2000, Fourmann et al., 2001). Some of these RGA mapped to LG N7 carrying race specific genes for resistance to *L. maculans* (Sillito et al., 2000). A *B. nigra* cDNA sequence, denoted *Lm1*, improved resistance to *L. maculans* in both cotyledons and leaves when it was expressed in transgenic oilseed rape (Wretblad et al., 2003). Identification of differential gene expression using microarray technology has been used to understand the interactions between *L. maculans* and resistant (*LepR3* gene)/susceptible host plants (Kaur et al., 2004). Work is in progress to clone the 'Crésor' resistance gene (I Parkin, unpubl.) and the *LepR3* gene introgressed from *B. rapa* var. *sylvestris* (Larkan et al., 2004). Isolation of *Rlm* genes that correspond to the avirulence genes that are being cloned in *L. maculans* (Kuhn et al., 2006) will make the *B. napus*/*L. maculans* pathosystem an excellent model system for studies of the molecular interactions between a host and its pathogen.

Conclusions

Our understanding of *L. maculans*/*Brassica* interactions has increased greatly in recent years with developments in genetic studies on both the pathogen and the host plant, and with increased knowledge of the distribution of avirulence alleles in *L. maculans* populations (Balesdent et al., 2005; Stachowiak et al., 2006). To understand the interactions, it is necessary to distinguish between two types of resistance; a qualitative resistance effective from the seedling to the adult plant stage (race specific resistance) and a quantitative adult-stage resistance that is controlled by many genes with small individual effects. Field resistance can be conferred by race specific major genes and/or by polygenes. Partial resistance in the field can be due either to a major gene on which the *L. maculans* population is partly virulent or to quantitative resistance. It is only through investigating the presence of race specific resistance gene(s) in the *B. napus* genotype tested and avirulence alleles in the *L. maculans* isolates used in controlled environment or field experiments that the type of resistance can be determined. Currently, it seems that the genes involved in race specific resistance and polygenic race non-specific resistance are distinct. However, mechanisms leading to quantitative resistance can be effective at different stages of epidemic development and may differ depending on the resistance source. A better understanding of the mechanisms underlying quantitative resistance would help our understanding of the relationships between quantitative and major resistance genes.

Leptosphaeria maculans populations have a very great potential to evolve to virulence under selection pressure exerted by race specific resistance genes and single resistance genes do not provide a durable resistance. This has been shown both in a field experiment using the *Jlm1*/*Rlm6* gene introgressed into *B. napus* from *B. juncea* (Brun et al., 2000) and in commercial crops for the *Rlm1* cultivars and the *LepR3* gene introgressed from *B. rapa* var. *sylvestris* (Li et al., 2003, 2005; Rouxel et al., 2003; Sprague et al., 2006). Polygenic resistance has generally been considered durable. This is supported by evidence for the commercial cultivar Jet Neuf. However, the polygenic resistance in some *B. napus* cultivars has become less effective with time. Sprague et al. (2006) reported that, after its release in 1993, 'Rainbow' maintained its Australia Blackleg Rating (ABR) for resistance at 6.5 until 2000 but it decreased to 5.5 in 2004 while the ABR of 'Ripper' decreased more rapidly, from 7.5 in 2000 to 5.0 in 2004. This is presumed to be the result of changes in virulence and aggressiveness (the ability to cause more severe disease) of the *L. maculans* population. It is difficult to test experimentally the hypothesis that aggressiveness has changed, since this requires multi-season, multi-location field testing of *B. napus* cultivars and continual monitoring of the pathogen population for aggressiveness and frequency of different pathotypes.

To maximise durability of resistance, it is necessary to identify as many different resistance genes as possible to diversify their use and establish strategies to manage them through genotype construction and deployment. To achieve this objective, there is a need to improve characterisation of the race specific resistance genes and QTL for non specific resistance to *L. maculans*.

Acknowledgements

French research was funded by the French Institut National de la Recherche Agronomique, CETIOM (Centre Technique Interprofessionnel des Oléagineux Métropolitains) and grants from the 'Société interprofessionelle des oléagineux, protéagineux et cultures textiles' (SIDO), the 'Agence de l' environnement et de la maitrise de l'énergie' (ADEME) and the Ministry of Agriculture. Part of the work was supported by the European project 'Integrated strategies the management of stem canker of oilseed rape in Europe' (IMASCORE). Funding for SR Rimmer was from Agriculture and Agri-Food Canada MII program, Blackleg Resistance Consortium, and Saskatchewan Agriculture and Food ADF program. Thanks are due to Derek Lydiate, Isobel Parkin and Andrew Sharpe for sharing information on alignment of different linkage maps. The Australian research was funded by the Grains Research and Development Corporation, the University of Melbourne and the state Departments of Primary Industries. We thank Bruce Fitt, Barbara Howlett, Maria Manzanares and Jon West for commenting on the paper during its preparation.

References

Ansan-Melayah D, Balesdent MH, Buée M and Rouxel T (1995) Genetic characterization of *AvrLm1*, the first avirulence gene of *Leptosphaeria maculans*. Phytopathology 85: 1525–1529

Ansan-Melayah D, Balesdent MH, Delourme R, Pilet ML, Tanguy X, Renard M and Rouxel T (1998) Genes for race-specific resistance against blackleg disease in *Brassica napus* L. Plant Breeding 117: 373–378

Badawy HMA, Hoppe HH and Koch E (1991) Differential reactions between the genus *Brassica* and aggressive single spore isolates of *Leptosphaeria maculans*. Journal of Phytopathology 131: 109–119

Balesdent MH, Attard A, Ansan-Melayah D, Delourme R, Renard M and Rouxel T (2001) Genetic control and host range of avirulence toward *Brassica napus* cultivars Quinta and Jet Neuf in *Leptosphaeria maculans*. Phytopathology 91: 70–76

Balesdent MH, Attard A, Kühn ML and Rouxel T (2002) New avirulence genes in the phytopathogenic fungus *Leptosphaeria maculans*. Phytopathology 92: 1122–1133

Balesdent MH, Barbetti MJ, Hua Li K, Sivasithamparam K, Gout L and Rouxel T (2005) Analysis of *Leptosphaeria maculans* race structure in a world-wide collection of isolates. Phytopathology 95: 1061–1071

Ballinger DJ and Salisbury PA (1996) Seedling and adult plant evaluation of race variability in *Leptosphaeria maculans* on *Brassica* species in Australia. Australian Journal of Experimental Agriculture 36: 485–488

Ballinger DJ, Salisbury PA and Kadkol GP (1991) Race variability in *Leptosphaeria maculans* and the implication for resistance breeding in Australia. Proceedings 8th International Rapeseed Congress, Saskatoon, Canada 1: 226–231

Bansal VK, Kharbanda PD, Stringam GR, Thiagarajah MR and Tewari JP (1994) A comparison of greenhouse and field screening methods for blackleg resistance in doubled haploid lines of *Brassica napus*. Plant Disease 78: 276–281

Barret P, Guérif J, Reynoird JP, Delourme R, Eber F, Renard M and Chèvre AM (1998) Selection of stable *Brassica napus–Brassica juncea* recombinant lines resistant to blackleg (*Leptosphaeria maculans*). 2. A 'to and fro' strategy to localise and characterise interspecific introgressions on the *B. napus* genome. Theoretical and Applied Genetics 96: 1097–1103

Bohman S, Wang M and Dixelius C (2002) *Arabidopsis thaliana*-derived resistance against *Leptosphaeria maculans* in a *Brassica napus* genomic background. Theoretical and Applied Genetics 105: 498–504

Brun H, Huteau V, Ermel M, Eber F, Chèvre AM and Renard M (2004) Field behaviour of oilseed rape genotypes carrying major resistance genes exposed to different *Leptosphaeria maculans* populations. International Organization for Biological Control Bulletin 27: 95–100

Brun H, Levivier S, Ruer D, Somda I, Chèvre AM and Renard M (2000) A field method for evaluating the potential durability of new resistance sources: application to the *Leptosphaeria maculans*/*Brassica napus* pathosystem. Phytopathology 90: 961–966

Cargeeg LA and Thurling N (1980) Contribution of host-pathogen interactions to the expression of the blackleg disease of spring rape (*Brassica napus* L.) caused by *Leptosphaeria maculans* (Desm.) Ces. et de Not. Euphytica 29: 465–476

Chen CY and Séguin-Swartz G (1999) Reaction of wild crucifers to *Leptosphaeria maculans*, the causal agent of blackleg of crucifers. Canadian Journal of Plant Pathology 21: 361–367

Chèvre AM, Barret P, Eber F, Dupuy P, Brun H, Tanguy X and Renard M (1997) Selection of stable *Brassica napus–Brassica juncea* recombinant lines resistant to blackleg (*Leptosphaeria maculans*). 1. Identification of molecular markers, chromosomal and genomic origin of the introgression. Theoretical and Applied Genetics 95: 1104–1111

Chèvre AM, Eber F, This P, Barret P, Tanguy X, Brun H, Delseny M and Renard M (1996) Characterization of *Brassica nigra* chromosomes and of blackleg resistance in *B. napus–B. nigra* addition lines. Plant Breeding 115: 113–118

Chèvre AM, dePonce Leon A, Jenczewski E, Eber F, Delourme R, Renard M and Brun H (2003) Introduction of blackleg resistance from *Brassica rapa* into *Brassica napus*. Proceedings 11th International Rapeseed Congress, Copenhagen, Denmark 1: 32–35

Crouch JH, Lewis BG and Mithen RF (1994) The effect of A-genome substitution on the resistance of *Brassica napus* to infection by *Leptosphaeria maculans*. Plant Breeding 112: 265–278

Delourme R, Pilet-Nayel ML, Archipiano M, Horvais R, Tanguy X, Rouxel T, Brun H, Renard M and Balesdent MH (2004) A cluster of major specific resistance genes to *Leptosphaeria maculans* in *Brassica napus*. Phytopathology 94: 578–583

Dion Y, Gugel RK, Rakow GFW, Séguin-Swartz G and Landry BS (1995) RFLP mapping of resistance to the blackleg disease [causal agent, *Leptosphaeria maculans* (Desm.) Ces. et de Not.] in canola (*Brassica napus* L.). Theoretical and Applied Genetics 91: 1190–1194

Dixelius C (1999) Inheritance of the resistance to *L. maculans* of *B. nigra* and *B. juncea* in near isogenic lines of *B. napus*. Plant Breeding 118: 151–156

Dixelius C and Wahlberg S (1999) Resistance to *L. maculans* is conserved in a specific region of *Brassica* B genome. Theoretical and Applied Genetics 99: 368–372

Ferreira ME, Rimmer SR, Williams PH and Osborn TC (1995) Mapping loci controlling *Brassica napus* resistance to *Leptosphaeria maculans* under different screening conditions. Phytopathology 85: 213–217

Ferreira ME, Williams PH and Osborn TC (1994) RFLP mapping of *Brassica napus* using doubled haploid lines. Theoretical and Applied Genetics 89: 615–621

Fourmann M, Charlot F, Froger N, Delourme R and Brunel D (2001) Expression, mapping, and genetic variability of *Brassica napus* disease resistance gene analogues. Genome 44: 1083–1099

Fristensky B, Balcerzak M, He D and Zhang P (1999) Expressed sequence tags from the defense response of *Brassica napus* to *Leptosphaeria maculans*. Molecular Plant Pathology On-Line: http://www.bspp.org.uk/mppol/1999/0301Fristensky

Joyeux A, Fortin MG, Mayerhofer R and Good AG (1999) Genetic mapping of plant disease resistance gene homologues using a minimal *Brassica napus* L. population. Genome 42: 735–743

Kaur J, Webster T, Felitti S, Batley J, Howlett BJ, Salisbury PA and Edwards D (2004) Microarray analysis of gene expression in canola in response to blackleg fungus (*Leptosphaeria maculans*) infection. Proceedings Joint Meeting 14th Crucifer Genetics Workshop, 4th ISHS Symposium on Brassicas, Brassica 2004, Daejeon, Korea, 196

Keri M, vanden Berg CGJ, McVetty PBE and Rimmer SR (1997) Inheritance of resistance to *Leptosphaeria maculans* in *Brassica juncea*. Phytopathology 87: 594–598

Khangura RK and Barbetti MJ (2001) Prevalence of blackleg (*Leptosphaeria maculans*) on canola (*Brassica napus*) in Western Australia. Australian Journal of Experimental Agriculture 41: 71–40

Kuhn ML, Gout L, Howlett BJ, Melayah D, Meyer M, Balesdent MH and Rouxel T (2006) Genetic linkage maps and genomic organization in *Leptosphaeria maculans*. European Journal of Plant Pathology 114: 17–31

Kuswinanti T, Sock J and Hoppe HH (1999) Virulence pattern of aggressive isolates of *Leptosphaeria maculans* on an extended set of *Brassica* differentials. Zeitschrift für Pflanzenkrankheiten und Pflanzenschutz 106: 12–20

Kutcher HR, vanden Berg CGJ and Rimmer SR (1993) Variation in pathogenicity of *Leptosphaeria maculans* on *Brassica* spp. based on cotyledon and stem reactions. Canadian Journal of Plant Pathology 15: 253–258

Larkan N, Nelson M, Yu F, Hammond C, Parkin IAP, Cowling WA, Lydiate DJ and Rimmer SR (2004) Mapping *LepR3*. using *Arabidopsis-Brassica* collinearity to breed blackleg-resistant canola. Proceedings Joint Meeting 14th Crucifer Genetics Workshop, 4th ISHS Symposium on Brassicas, Brassica 2004, Daejeon, Korea: 179

Li CX and Cowling WA (2003) Identification of a single dominant allele for resistance to blackleg in *Brassica napus* 'Surpass 400'. Plant Breeding 122: 485–488

Li H, Barbetti MJ and Sivasithamparam K (2005) Hazard from reliance on cruciferous hosts as source of major gene-based resistance for managing blackleg (*Leptosphaeria maculans*) disease. Field Crops Research 91: 185–198

Li H, Sivasithamparam K and Barbetti MJ (2003) Breakdown of a *Brassica rapa* subsp. *sylvestris* single dominant blackleg resistance gene in *Brassica napus* rapeseed by *Leptosphaeria maculans* field isolates in Australia. Plant Disease 87: 752

Lombard V and Delourme R (2001) A consensus linkage map for rapeseed (*Brassica napus* L.): construction and integration of three individual maps from DH populations. Theoretical and Applied Genetics 103: 491–507

Mahuku GS, Hall R and Goodwin PH (1996) Co-infection and induction of systemic acquired resistance by weakly and highly virulent isolates of *Leptosphaeria maculans* in oilseed rape. Physiological and Molecular Plant Pathology 49: 61–72

Marcroft SJ, Sprague SJ, Pymer SJ, Salisbury PA and Howlett BJ (2003) Factors affecting production of inoculum of the blackleg fungus *Leptosphaeria maculans* in south-eastern Australia. Australian Journal of Experimental Agriculture 43: 1231–1236

Marcroft SJ, Sprague SJ, Salisbury PA and Howlett BJ (2004) Potential for using host-resistance to reduce production of pseudothecia and ascospores of *Leptosphaeria maculans*, the blackleg pathogen of *Brassica napus*. Plant Pathology 53: 468–474

Mayerhofer R, Bansal VK, Thiagarajah GR, Stringam GR and Good AG (1997) Molecular mapping of resistance to *Leptosphaeria maculans* in Australian cultivars of *Brassica napus*. Genome 40: 294–301

Mayerhofer R, Wilde K, Mayerhofer M, Lydiate D, Bansal V, Good A, and Parkin I (2005) Complexities of chromosomes landing in a highly duplicated genome: Towards map based cloning of a gene controlling blackleg resisitance in *Brassica napus*. Genetics 2005; published ahead of print on September 2, 2005 as doi: 10.1534/genetics.105.049098

McNabb WM, vanden Berg CGJ and Rimmer SR (1993) Comparison of inoculation methods for selection of plant resistant to *Leptosphaeria maculans* in *Brassica napus*. Canadian Journal of Plant Science 73: 1199–1207

Mengistu A, Rimmer SR, Koch E and Williams PH (1991) Pathogenicity grouping of isolates of *Leptosphaeria maculans* on *Brassica napus* cultivars and their disease reaction profiles on rapid cycling *Brassicas*. Plant Disease 75: 1279–1282

Mithen RF and Lewis BG (1988) Resistance to *Leptosphaeria maculans* in hybrids of *Brassica oleracea* and *Brassica insularis*. Journal of Phytopathology 123: 253–258

Mithen RF, Lewis BG, Heaney RK and Fenwick GR (1987) Resistance of leaves of *Brassica* species to *Leptosphaeria maculans*. Transactions of the British Mycological Society 88: 525–531

Newman PL and Bailey DJ (1987) Screening for resistance to canker (*Leptosphaeria maculans*) in winter oilseed rape (*Brassica napus* ssp. *oleifera*). Plant Pathology 36: 346–354

Ogbonnaya FC, Halloran GM, Marcroft SJ, Pang ECK and Gororo N (2003) Progress in the utilisation of *Brassica nigra* in breeding for resistance to blackleg (*Leptosphaeria maculans*). Proceedings 11th International Rapeseed Congress, Copenhagen, Denmark 1: 39–41

Osborn TC, Butruille DV, Sharpe AG, Pickering KJ, Parkin IAP, Parker JS and Lydiate DJ (2003) Detection and effects of a homeologous reciprocal transposition in *Brassica napus*. Genetics 165: 1569–1577

Pang ECK and Halloran GM (1996a) The genetics of adult-plant blackleg (*Leptosphaeria maculans*) resistance from *Brassica juncea* in *B. napus*. Theoretical and Applied Genetics 92: 382–387

Pang ECK and Halloran GM (1996b) The genetics of blackleg [*Leptosphaeria maculans* (Desm.) Ces. et De Not.] resistance in rapeseed (*Brassica napus* L.). II. Seedling and adult-plant resistance as quantitative traits. Theoretical and Applied Genetics 93: 941–949

Parkin IAP, Sharpe AG, Keith DJ and Lydiate DJ (1995) Identification of the A and C genomes of amphidiploid *Brassica napus* (oilseed rape). Genome 38: 1122–1131

Pilet ML, Delourme R, Foisset N and Renard M (1998) Identification of loci contributing to quantitative field resistance to blackleg disease, causal agent *Leptosphaeria maculans* (Desm.) Ces. et de Not., in winter rapeseed (*Brassica napus* L.). Theoretical and Applied Genetics 96: 23–30

Pilet ML, Duplan G, Archipiano M, Barret P, Baron C, Horvais R, Tanguy X, Lucas MO, Renard M and Delourme R (2001) Stability of QTL for field resistance to blackleg across two genetic backgrounds in oilseed rape. Crop Science 41: 197–205

Plieske J, Struss D and Röbbelen G (1998) Inheritance of resistance derived from the B-genome of *Brassica* against *Phoma lingam* in rapeseed and the development of molecular markers. Theoretical and Applied Genetics 97: 929–936

Rimmer SR (2006) Resistance genes to *Leptosphaeria maculans* in *Brassica napus*. Canadian Journal of Plant Pathology: (in press)

Rimmer SR and vanden Berg CGJ (1992) Resistance of oilseed *Brassica* spp. to blackleg caused by *Leptosphaeria maculans*. Canadian Journal of Plant Pathology 14: 56–66

Rouxel T, Willner E, Coudard L and Balesdent MH (2003) Screening and identification of resistance to *Leptosphaeria maculans* (stem canker) in *Brassica napus* accessions. Euphytica 133: 219–231

Roy NN (1978) A study on disease variation in the populations of an interspecific cross of *Brassica juncea* L. x *Brassica napus* L. Euphytica 27: 145–149

Roy NN (1984) Interspecific transfer of *Brassica juncea*-type high blackleg resistance to *Brassica napus*. Euphytica 33: 295–303

Roy NN, Fisher HM and Tarr A (1983) Wesbrook – a new prime variety of rapeseed. Proceedings Fourth Australian Rapeseed Agronomists and Breeders Workshop, Lyndoch, 4pp

Saal B, Brun H, Glais I and Struss D (2004) Identification of a *Brassica juncea*-derived recessive gene conferring resistance to *Leptosphaeria maculans* in oilseed rape. Plant Breeding 123: 505–511

Sacristan MD and Gerdemann M (1986) Different behavior of *Brassica juncea* and *Brassica carinata* as sources of *Phoma lingam* resistance in experiments of interspecific transfer to *Brassica napus*. Plant Breeding 97: 304–314

Salisbury PA, Ballinger DJ, Wratten N, Plummer KM and Howlett BJ (1995) Blackleg disease on oilseed *Brassica* in Australia: a review. Australian Journal of Experimental Agriculture 35: 665–672

Salisbury PA and Wratten N (1999). *Brassica napus* breeding. In: Salisbury PA, Potter T, McDonald G and Green AG (eds) Canola in Australia: The First Thirty Years (pp 29–35) Canberra, Australia

Shoemaker RA and Brun H (2001) The teleomorph of the weakly aggressive segregate of *Leptosphaeria maculans*. Canadian Journal of Botany 79: 412–419

Sillito D, Parkin IAP, Mayerhofer R, Lydiate DJ and Good AG (2000) *Arabidopsis thaliana*: a source of candidate disease-resistance genes for *Brassica napus*. Genome 43: 452–460

Sjödin C and Glimelius K (1989) Transfer of resistance against *Phoma lingam* to *B. napus* by asymetric somatic hybridization combined with toxin selection. Theoretical and Applied Genetics 78: 513–520

Snowdon RJ, Winter H, Diestel A and Sacristan MD (2000) Development and characterization of *Brassica napus-Sinapis arvensis* addition lines exhibiting resistance to *Leptosphaeria maculans*. Theoretical and Applied Genetics 101: 1008–1014

Sprague SJ, Balesdent MH, Brun H, Hayden HL, Marcroft S, Pinochet X, Rouxel T and Howlett BJ (2006) Major gene resistance of *Brassica napus* (oilseed rape) is overcome by changes in virulence of populations of *Leptosphaeria maculans* in France and Australia. European Journal of Plant Pathology 114: 33–40

Stachowiak A, Olechnowicz J, Jedryczka M, Rouxel T, Balesdent MH, Happstadius I, Gladders P, Latunde-Dada AO and Evans N (2006) Frequency of avirulence alleles in field populations of *Leptosphaeria maculans* in Europe. European Journal of Phytopathology 114: 67–75

Williams PH and Delwiche PA (1979) Screening for resistance to blackleg of crucifers in the seedling stage. Proceedings Eucarpia Conference, Breeding of Cruciferous Crops, Wageningen, Netherlands: 164–170

Winter H, Gaertig S, Diestel A and Sacristan MD (1999) Blackleg resistance of different origin transferred into Brassica napus. Proceedings of the 10th International Rapeseed Congress, Canberra, Australia (http://www.regional.org.au/papers/index.htm)

Wretblad S, Bohman S and Dixelius C (2003) Over-expression of a *Brassica nigra* cDNA gives enhanced resistance to *Leptosphaeria maculans* in *B. napus*. Molecular Plant-Microbe Interactions 16: 477–484

Yu F, Lydiate DJ and Rimmer SR (2004) Identification and mapping of a third blackleg resistance locus in *Brassica napus* derived from *B. rapa subsp. sylvestris*. Plant, Animal & Microbe Genomes XII Conference, San Diego, California, USA

Yu F, Lydiate DJ and Rimmer SR (2005) Identification of two novel genes for blackleg resistance in *Brassica napus*. Theoretical and Applied Genetics 110: 969–979

Zhu B and Rimmer SR (2003) Inheritance of resistance to *Leptosphaeria maculans* in two accessions of *Brassica napus*. Canadian Journal of Plant Pathology 25: 98–103

Zhu JS, Struss D and Röbbelen G (1993) Studies on resistance to *Phoma lingam* in *Brassica napus-Brassica nigra* addition lines. Plant Breeding 111: 192–197

European Journal of Plant Pathology (2006) 114: 53–65
DOI 10.1007/s10658-005-2104-0

A large-scale survey of races of *Leptosphaeria maculans* occurring on oilseed rape in France

Marie-Hélène Balesdent[1,*], Karine Louvard[1,2], Xavier Pinochet[2] and Thierry Rouxel[1]
[1]*Unité PMDV, INRA, Route de Saint Cyr, 78026, Versailles Cedex, France;* [2]*CETIOM (Centre Technique des Oléagineux Métropolitains), BP4, 78850, Thiverval-Grignon, France; *Author for correspondence (Phone: +33-1-30-833229; Fax: +33-1-30-833195; E-mail: mhb@versailles.inra.fr)*

Accepted 12 August 2005

Key words: avirulence gene, blackleg disease, gene-for-gene relationship, *Phoma lingam*, resistance

Abstract

Nine avirulence genes (*AvrLm1–AvrLm9*) were identified in *Leptosphaeria maculans*, the causal agent of stem canker of oilseed rape (OSR), combinations of which could theoretically generate up to 512 different races of the fungus. *L. maculans* displays a high evolutionary potential to adapt to novel resistance genes as illustrated by the *Rlm1* breakdown in France, where virulent populations became prevalent within three growing seasons. An improved knowledge of the race structure of the fungal population is therefore needed to ensure a better use of available major resistance genes. The objective of this study was to characterise the *L. maculans* population structure in France using a large-scale, rationalised sample of isolates. Experimental fields, planted with "trap plants" harbouring no major resistance gene, were sown at 20 locations. Single-pycnidium isolates were collected from leaf lesions that developed in early autumn and 1797 isolates were genotyped at Avr loci. The frequency of *AvrLm6* and *AvrLm7* was higher than 99%, whereas *avrLm2* and *avrLm9* alleles were fixed in the population. *AvrLm1*, *AvrLm4*, *AvrLm5* and *AvrLm8* were polymorphic. *AvrLm3* isolates were detected at a very low frequency (less than 1%). Only 11 races were identified in France, with one race prevalent, namely Av5-6-7-(8) (i.e. virulent on *Rlm1*, *Rlm2*, *Rlm3*, *Rlm4* and *Rlm9*), representing around 65% of the population. Disparities between the locations sampled were evident at all scales analysed. Some virulent races, such as those harbouring *avrLm5*, were present before the introduction of the corresponding resistance gene in the commercial OSR crop.

Abbreviations: Avr – avirulence; cv – cultivar; OSR – oilseed rape; R – resistance

Introduction

The dothideomycete *Leptosphaeria maculans* (anamorph *Phoma lingam*), the causal agent of phoma stem canker disease of oilseed rape (OSR) (winter *Brassica napus*) and canola (spring *B. napus* and *B. rapa*), is the most economically important disease of oilseed Brassicas worldwide. Breeding for "field resistance" to *L. maculans* in the past often gave rise to cvs harbouring single major gene resistance (*Rlm* genes) that were selected unintentionally due to their complete effectiveness when challenged by populations of the pathogen harbouring the corresponding avirulence (*AvrLm*) allele (Ansan Melayah et al., 1997; Pinochet et al., 2003; Rouxel et al., 2003a). Most currently registered winter OSR cvs in France harbour at least one *Rlm* gene (Pinochet et al., 2003, 2004; Balesdent and Pinochet unpublished data). The life cycle of the pathogen involves annual sexual recombination and large-scale dispersal of ascospores (West et al., 2001), which predisposes *L. maculans* to have a high evolutionary potential to adapt to novel resistance genes (McDonald and Linde, 2002). This was illustrated by the case of *Rlm1* in France, where cvs harbouring this gene became popular and accounted for more than 40% of the French OSR production area (Rouxel et al., 2003a). This was associated with a concurrent increase in the proportion of virulent individuals within the population of *L. maculans* within three growing seasons (Rouxel et al., 2003a). This study, along with experimental data which indicate that the novel *Rlm6* gene (not commercially used to date in Europe) can also be overcome within three growing seasons when subjected to artificially increased inoculum pressure (Brun et al., 2000), and the current breakdown of the newly-introduced "Surpass 400" resistance in Australia (Li et al., 2003; Sprague et al., 2006), illustrate the need to monitor the race structure of *L. maculans* populations. Such information will allow more efficient use of available sources of major gene resistance (R) to prevent pathogen adaptation and the spread of novel virulent isolates to regions where novel resistances have not been used.

Although based on a large collection (1011) of French isolates, the study by Rouxel et al. (2003a) was insufficient to provide an extensive survey of the current structure of *L. maculans* populations in France since (i) it included samples taken between 1994 and 2000 at only three sampling locations, (ii) only three Avr genes were analysed, *AvrLm1*, *AvrLm2* and Av*rlm4*, whereas nine Avr genes have currently been identified (Balesdent et al., 2002, 2005; Delourme et al., 2004), (iii) part of the collection originated from unknown (or uncharacterised in terms of R genes) plant genotypes, which might exclude some races avirulent on these genotypes, resulting in a sampling bias caused by selection of the corresponding virulent alleles. For example, the *Rlm2* gene was extremely popular in French cvs (Rouxel et al., 2003a, b) and it was impossible to ascertain that the corresponding avirulent allele, *AvrLm2*, was absent in *L. maculans* populations in France.

A recent analysis of the race composition of an international *L. maculans* isolate collection for all nine identified Avr genes indicated that there was a low diversity of races in Europe, with only 8 out of 512 possible races identified, and important differences between continents in race structure (Balesdent et al., 2005). However, only historical isolates were present in this collection, and the information on the plant genotype from which the isolates originated was often absent.

In this context, the objectives of the present study were (i) to extensively investigate the *L. maculans* population structure in France, including all nine *AvrLm* genes known to date, based on a rational sampling scheme, (ii) to evaluate the putative regional differences in race structure and (iii) to relate them to the history of OSR production in the surrounding region. It was also envisaged that such a large survey would allow us to evaluate which (and where) *Rlm* genes were still effective against a significant part of the *L. maculans* population, and if some virulent races were already present before the introduction of unused novel resistance genes. For this purpose, large-scale sampling of the *L. maculans* population of France was done, on a country-wide basis from a network of similar experimental fields planted with "trap plants" (known to harbour no major resistance genes). These crops were grown by a network of OSR breeding companies, public research institutes, cultivar registration bodies and advisory bodies. The resulting collection of isolates was analysed for the occurrence of *AvrLm* alleles corresponding to *Rlm* genes which had been used commercially in the past (*Rlm2*, *Rlm3*, *Rlm4*, *Rlm9*), more recently introduced commercially (*Rlm1*) or not yet been used commercially [*Rlm7* from *B. napus*, *Rlm5* and *Rlm6* from *B. juncea* and *Rlm8* from *B. rapa* (Balesdent et al., 2002)]. Finally, control plant genotypes, each having different known specific resistance genes, were also included in each trial, to relate the *L. maculans* population structure to effectiveness of specific resistance genes in winter OSR crops.

Materials and methods

Experimental fields

Seventeen experiments were sown in August–September 2000, and three experiments were sown in August–September 2001 to cover important OSR production areas that were not included in the 2000/2001 survey (Figure 1). In 2000/2001, the experimental design comprised 10 cvs sown in four replicates at each site (Table 1). Each plot (one replicate of one cv. on one site) consisted of four lines of minimum length 2 m. Cultivar Drakkar, a

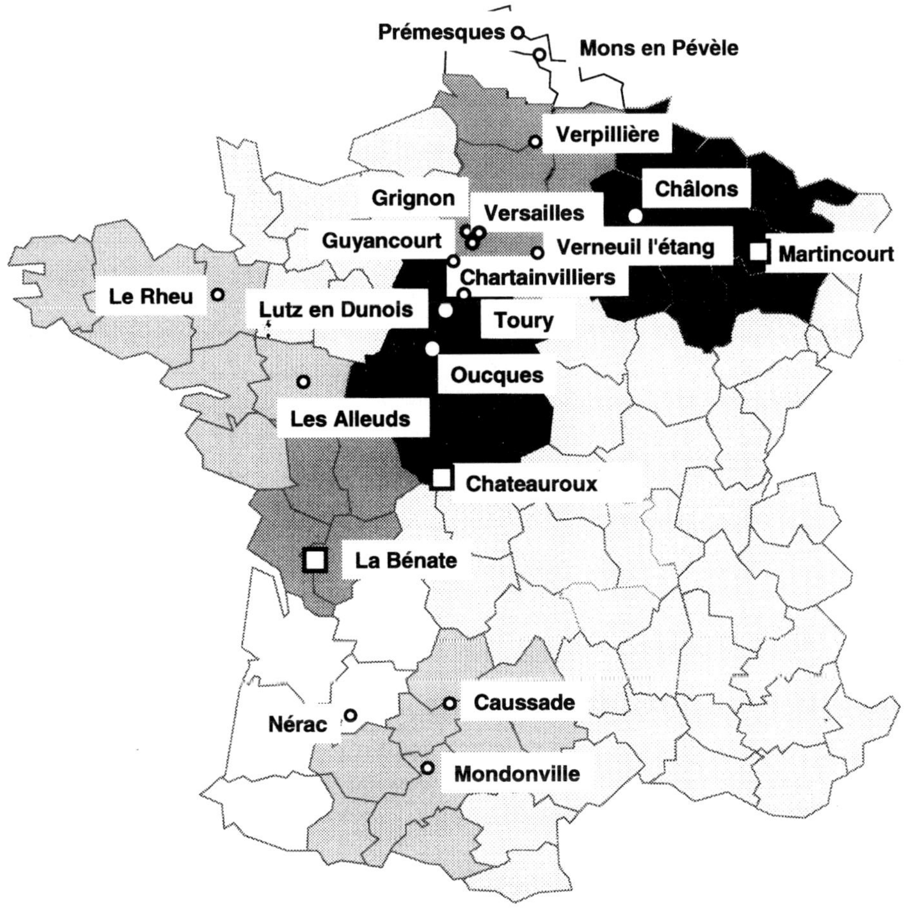

Figure 1. Location of the 20 OSR experimental sites sampled across France (○, sites sampled in Autumn 2000 □, sites sampled in Autumn; 2001). Regions in white, less than 20,000 ha of OSR cultivated in this region; regions in pale grey, between 20,000 and 50,000 ha; regions in medium grey, between 50,000 and 100,000 ha; regions in black, from 100,000 to 250,000 ha. Other regions were not sampled.

spring genotype known to possess no *Rlm* gene (provided by INRA-APBV, Le Rheu) was used as a "trap plant", because it exerted no direct selection on the *L. maculans* population. Cv. Surpass 400, possessing a *B. rapa* subsp. *sylvestris* major gene termed *LepR3* (Li and Cowling, 2003; Yu et al., 2004) and line 23-1-1 (*Rlm7*) were assessed as potential novel resistance sources for the French conditions, as neither had been previously used commercially in France. The other cvs chosen were commercial cultivars with or without *Rlm1*, *Rlm2*, *Rlm4* or *Rlm9* (Table 1). Jet Neuf (*Rlm4*) was also grown as a reference cultivar. In 2001/2002, the experiment included only cv. Drakkar, with four replicates at each site. No additional *L. maculans* inoculum (e.g. infected debris) was used at the beginning of the growing season. Sowing date and agricultural practices were as close as possible to what was usual in that particular region.

The incidence and severity of phoma leaf spots and stem canker were recorded in 2000/2001. Incidence (i.e. percentage of plants with at least one leaf lesion) and severity of phoma leaf spots were recorded in autumn at 11 sites by observing 60 plants per block (6 row-lengths of 10 plants each). Severity of the leaf spotting was recorded on the same set of plants by placing each plant into one of four classes: 0, no leaf symptom; 1, <5 spots per plant; 2, 5–10 leaf spots per plant; 3, >10 leaf spots per plant. These data were used to calculate a leaf disease index (LDI) according to the formula: $LDI = \Sigma_i (n_i c_i)/N$, where n_i is the number of plants in class i, c_i is a correction coefficient score for class i (1 for class 1; 3 for class 2; 5 for

Table 1. Characteristics of the OSR genotypes sown at 17 experimental sites across France in 2000/2001

Cultivar/line	Resistance gene(s)	Provider	Type of cv.	Phoma leaf spot index[a]	Stem canker index[b]
Drakkar	None	INRA-APBV	Spring	2.00 ± 1.50[c]	nd
Surpass 400	*LepR3*	Advanta seeds	Spring	1.01 B	nd
Pollen	*Rlm4*	Momont Hennette	Winter	1.24 B	2.72 D
Bristol	*Rlm2, Rlm9*	Monsanto SAS	Winter	1.27 B	4.53 A
Columbus	*Rlm1, Rlm3*	Monsanto SAS	Winter	1.26 B	2.26 E
Vivol	*Rlm1, Rlm3*	Monsanto SAS	Winter	1.12 B	3.54 C
Mohican	*Rlm9*	Rustica prograin génétique	Winter	1.38 AB	3.97 B
Jet Neuf	*Rlm4*	INRA-PMDV	Winter	1.32 B	1.76 F
Goéland	*Rlm9*	Momont Hennette	Winter	1.84 A	3.21 C
23-1-1	*Rlm7*	INRA-PMDV	Winter	0.45 C	0.58 G

[a]Values are mean indices (0–5 scale) from 11 experimental sites for 2000–2001, for which complete phoma leaf spot incidence and severity data were available. Cvs followed by the same uppercase letter are not significantly different at the 5% level (Bonferroni test).
[b]Values are mean indices (0–9 scale) from 14 sites with complete stem canker (G2 index) ratings (2000–2001). Cvs followed by the same uppercase letter are not significantly different at the 5% level (Bonferroni test). nd, not determined (spring-type cvs).
[c]Cv. Drakkar (only assessed at a few sites) was not included in the statistical analysis.

class 3), and N is the total number of plants. LDI values therefore ranged between 0 (no disease) and 5 (all plants severely affected). At the Versailles site, this index correlated well with the number of phoma leaf spots ($R^2 = 0.98$, data not shown). Severity of stem canker was assessed in mid-June to early July using the G2 stem canker index, ranging from 0 (no disease) to 9 (all plants lodged) (Pierre and Regnault, 1982; Aubertot et al., 2004).

Isolation and culture of L. maculans

When phoma leaf spot symptoms appeared (between October and December), 25 leaves with at least one leaf lesion each were collected per block of cv. Drakkar, washed, dried, and sent to INRA-PMDV for immediate isolation of *L. maculans*. Single-pycnidium isolates were collected from one lesion per affected leaf (West et al., 2002). All fungal cultures, for sporulation or long-term storage, were processed as described by Ansan-Melayah et al. (1995).

The plant differential set

A differential set, comprising genetically fixed cultivars or lines possessing only a few *Rlm* genes, was used to identify races of *L. maculans* (Balesdent et al., 2005). The differential set consisted of cvs/lines Westar (no R genes, susceptible control), Columbus (*Rlm1, Rlm3*), Bristol (*Rlm2, Rlm9*), 22-1-1 (*Rlm3*), Jet Neuf (*Rlm4*), 150-2-1 (*B. juncea* line, *Rlm5*, not characterised at the *Rlm9* locus), Falcon-MX (*Rlm4, Rlm6*, not characterised at the *Rlm9* locus) or Samouraï-MX (*Rlm1, Rlm6*, not characterised at the *Rlm9* locus) or Darmor-MX (*Rlm6*, not characterised at the *Rlm9* locus), 23-1-1 (*Rlm7*), 156-2-1 (*B. rapa* line, *Rlm8*, not characterised at the *Rlm9* locus) and Goéland (*Rlm9*). The commercial cvs Columbus and Bristol were provided by Monsanto SAS (Boissay, France) and cv. Goéland by Momont Hennette (Mons-en-Pévèle, France). Samouraï-MX, Darmor-MX and Falcon-MX were provided by M. Renard (INRA-APBV, Le Rheu, France). All other genotypes were maintained as described by Balesdent et al. (2002).

Pathogenicity tests and race terminology

Isolates were inoculated onto cotyledons of the plant differential set (Balesdent et al., 2001, 2005). Each isolate/line interaction was recorded as compatible (>80% susceptibility symptoms, the isolate is virulent [*avrLm*]) or incompatible (>80% resistance responses, the isolate is avirulent [*AvrLm*]) (Balesdent et al., 2001). Combining the results of all interaction phenotypes for a given isolate on the whole differential set made it possible to identify each Avr allele for all Avr loci. For instance, an isolate which was virulent on the line 22-1-1 (*Rlm3*) but avirulent on Columbus (*Rlm1, Rlm3*) was classified as being avirulent at the AvrLm1 locus and virulent at the AvrLm3 locus. The isolates were then classified into races, based on their pattern of avirulence alleles as described previously (Balesdent et al., 2005). For instance, the race

Av1-2-4-7 is composed of isolates possessing the Avr alleles *AvrLm1*, *AvrLm2*, *AvrLm4* and *AvrLm7*. Figures within brackets correspond to Avr loci for which genotyping was not possible.

Data analysis

Phoma leaf spot incidence, phoma leaf spot severity, stem canker severity (G2 rating), and frequencies of each Avr allele were analysed by analysis of variance, with the sites and/or the plant genotypes as sources of variation, and blocks as replicates. The plant genotypes or the sites were classified using the Bonferroni test ($\alpha = 0.05$). Statistical analyses were performed using XLStat 7.5 software. Populations (isolates obtained from a given site) were compared for frequency of each avirulent allele and for complexity of the isolates (number of virulences per isolate). In addition, two indices were used to analyse populations diversity (Pinon and Frey, 1997): the Margalef index, which measures the richness in species (here, in races) of a population, and the Simpson index of richness, that also takes into account the evenness of races within each population:

Margalef index:
$D_{Mg} = (S-1)/\log_e N$, with S = number of races and N = total number of isolates, at a site.

Simpson index:
$D_S = \Sigma\ [(n_i(n_i-1))/N(N-1)]$, with n_i = number of isolates of race i and N = number of isolates, at a site. The Simpson diversity index was calculated as $SD = 1 - D_S$.

Results

Disease severity in the 2000–2001 growing season

Incidence and severity of phoma leaf spot and stem canker data indicated that *L. maculans* inoculum concentration differed between sites in the 2000–2001 growing season. Incidence (% plants affected) of phoma leaf spot ranged from 12% (Mons) to 99% (Grignon), whereas phoma leaf spot severity ranged from 0.31 (Versailles) to 3.5 (Grignon) on a 0–5 scale. Analysis of variance revealed significant differences between cultivars or lines and between sites for both disease incidence and severity, but no correlation was observed between the variable measured and either the date of assessment or the geographical location (data not shown). For example, the two sites with the lowest and highest values for disease severity were located within 20 km of each other, with symptoms assessed on dates only 15 days apart. For both phoma leaf spot incidence and severity, whatever the experimental site, the line possessing *Rlm7* (23-1-1) was significantly less affected than other cvs (Table 1, and data not shown). Surpass 400 developed some leaf spotting in autumn. Although Surpass 400 had the second lowest leaf disease indices, it did not differ significantly from most winter-type cvs (Table 1).

Stem canker severity ranged from 0.6 to 4.5 on a 0–9 scale, depending on the cv (Table 1). Although the stem canker severity was low by comparison with other seasons, there were significant differences between sites ($F = 66.3$, df = 13, $p < 0.0001$) and between lines ($F = 189.7$, df = 7, $p < 0.0001$). Line 23-1-1 was always the most resistant line, whatever the experimental site (Table 1 and data not shown). Cultivars Jet Neuf and Columbus were usually ranked as the most resistant after 23-1-1, whereas Bristol and Mohican were always ranked as most susceptible (Table 1 and data not shown). Bristol was the most susceptible cv in 8 out of 14 sites. Stem canker resistance of cv Surpass 400 could not be evaluated, as most of the plants of this spring cv. did not survive the winter.

Establishment of the collection of isolates

Affected leaves of cv Drakkar collected between October and December 2000 (17 sites), or October and December 2001 (3 sites), were sampled to collect a total of 1988 single-pycnidium isolates. Except for two sites, a total of 73–120 isolates was obtained from each site. Of these 1988 isolates, 1787 were characterised for their pattern of Avr alleles (no more than 100 isolates per site were analysed) (Table 2). Only eight out of the nine Avr loci could be characterised for all sites, due to problems in multiplying the *B. rapa* genotype harbouring *Rlm8* (156-2-1). The occurrence of *AvrLm8* could therefore be fully characterised for only one site (Châteauroux, in the Central area), and some data were obtained for a second site (Le Rheu).

Table 2. Characteristics of the race structure of *Leptosphaeria maculans* populations at 20 experimental sites in France

Site	Region[a]	No. isolates analysed	No. races[b]	Margalef Index	No. virulence alleles per isolate	% isolates avirulent for polymorphic Avr loci		
						AvrLm1	*AvrLm4*	*AvrLm5*
Chartainvilliers	C	100	7	1.30	4.76	12.00	19.00	93.00
Chateauroux	C	93	6	1.10	5.00	11.83	4.30	89.25
Lutz	C	100	8	1.52	4.73	30.00	5.00	91.00
Oucques	C	100	8	1.52	4.88	20.00	11.00	76.00
Toury	C	100	8	1.52	4.75	16.00	13.00	97.00
Mean for Central region			*7.40*	*1.39*	*4.82*	*17.97*	*10.46*	*89.25*
Châlons	E	89	8	1.56	5.06	31.46	5.62	56.18
Martincourt	E	75	5	0.93	5.03	2.67	5.33	89.33
Mean for Eastern region			*6.50*	*1.24*	*5.04*	*17.06*	*5.48*	*72.76*
Mons en P.	N	43	4	0.80	4.72	18.60	9.30	100.00
Prémesques	N	98	8	1.53	4.79	33.67	10.20	77.55
Mean for North region			*6.00*	*1.16*	*4.75*	*26.14*	*9.75*	*88.78*
Grignon	NC	100	5	0.87	4.88	6.00	11.00	95.00
Guyancourt	NC	50	6	1.28	5.08	4.00	14.00	74.00
Verneuil	NC	100	5	0.87	4.68	25.00	8.00	99.00
Verpilliere	NC	100	7	1.30	4.88	21.00	13.00	78.00
Versailles	NC	87	6	1.12	4.84	13.79	19.54	82.76
Mean for North-Central region			*5.80*	*1.09*	*4.87*	*13.96*	*13.11*	*85.75*
La Bénate	W	100	3	0.43	4.87	6.00	7.00	100.00
Le Rheu	W	100	8	1.52	4.62	39.00	8.00	91.00
Les Alleuds	W	73	6	1.16	4.85	27.40	6.85	80.82
Mean for Western region			*5.67*	*1.04*	*4.78*	*24.13*	*7.28*	*90.61*
Caussade	S	79	4	0.69	5.04	22.78	0.00	73.42
Mondonville	S	100	5	0.87	4.87	21.00	2.00	90.00
Nérac	S	100	4	0.65	5.37	20.00	0.00	43.00
Mean for South region			*4.33*	*0.74*	*5.09*	*21.27*	*0.67*	*68.81*
Mean for all sites[c]				*1.13* (±0.16)	*4.88* (±0.52)	*19.1* (±4.79)	*8.6* (±2.54)	*83.8* (±6.89)

[a]C, central region; E, eastern region; NC, north central region; W, western region; S, south region; for a guide to the location of the sites, see Figure 1.
[b]The number of races is calculated on the basis of Avr combinations for *AvrLm1–AvrLm7* and *AvrLm9*; data for *AvrLm8* are excluded because there were too few data available at this locus.
[c]Values between parentheses indicate the confidence interval at a 5 % level (Student test).

Avirulence alleles occurring in France

Two avirulence alleles, *AvrLm6* and *AvrLm7*, were present in more than 99% of the French *L. maculans* population (Figure 2). Only one isolate with *avrLm6* and one isolate with *avrLm7* were identified out of the total of 1787 isolates. These two isolates came from an experimental field located close to a stem canker nursery at one of the oldest OSR breeding sites in France. Two virulence alleles were fixed in the populations, as all isolates were virulent at the AvrLm2 and AvrLm9 loci. Only three isolates, from three different sites, were avirulent at the AvrLm3 locus (Figure 2). *AvrLm1* and *AvrLm4* avirulence alleles were harboured by a small proportion of the population (19.6 and 8.6%, respectively). In contrast, the majority of the isolates were avirulent at the AvrLm5 and AvrLm8 loci (83.8 and 68.4%, respectively) (Figure 2).

Regional differences for individual avirulence alleles

Site to site differences were observed for all three avirulence alleles which were polymorphic in the French population, i.e. *AvrLm1*, *AvrLm4* and *AvrLm5*, as *AvrLm8* could not be investigated nationally (Table 2). Depending on the site, *AvrLm1* was present in 2.7–39% of the isolates, with significant differences between sites ($F = 6.54$, $p < 0.0001$). This Avr allele was present at high frequencies in the Northern and Western areas where it was present at frequencies of up to 39% of the isolates. Regions where OSR is not one of the

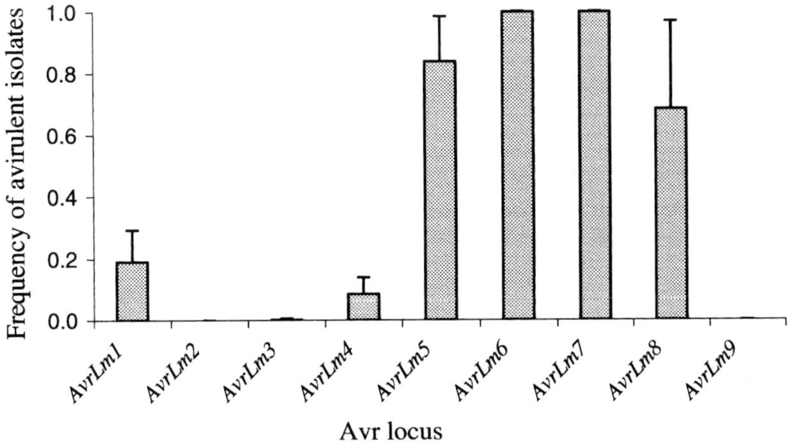

Figure 2. Frequency of avirulent (*AvrLm*) alleles for the nine Avr loci AvrLm1–AvrLm9 for isolates of *L. maculans* sampled at 20 experimental sites across France. Values are means over the 20 sites in Figure 1. Vertical lines are the standard deviation. A total of 1787 *Leptosphaeria maculans* isolates were analysed at each locus, except for *AvrLm8* (129 isolates).

main crops (Southern, Western and Northern areas, Figure 1) had the highest frequencies of isolates harbouring *AvrLm1* (Table 2). Within the main OSR growing areas, there were large differences between individual sites, with a few sites where the *AvrLm1* allele was almost absent (e.g. Martincourt in the Eastern area) (Table 2). Significant differences between sites ($F = 7.283$, $p < 0.0001$) were also observed within a region. For example, for the seven sites within a 100-km radius south of Paris (i.e. Guyancourt, Versailles, Grignon, Verneuil, Lutz, Toury and Chartainvilliers; Figure 1), the frequency of *AvrLm1* allele in the populations ranged between 4 and 30%. Due to this variation between sites within a region, no significant differences between regions were observed in frequency of *AvrLm1* ($F = 0.569$, $p = 0.723$).

Significant differences between sites were also observed for *AvrLm4* ($F = 3.225$, $p < 0.0001$). Depending on the site, the *AvrLm4* allele was present in 0 19.5% of the isolates (Table 2). The range of variation between sites within a region was small. In the Southern area, the allele was rare, and it was absent from two sites. In the North-Central area and three sites located in the North of the Central area (Oucques, Toury and Chartainvilliers) the frequency was >10% of the population (Table 2). A significant difference ($p = 0.0012$) between these populations and those from the Southern region was observed. The frequency of *AvrLm5* varied between sites, ranging from 43 to 100%, with significant differences between individual sites ($F = 8.1$, $p < 0.0001$), even when comparing the seven neighbouring sites around Paris ($F = 3.831$, $p = 0.01$). Consequently no evidence for regional differences was observed ($F = 1.255$, $p = 0.336$).

Race structure at the national and regional scales

As most *AvrLm8* data were missing, race structure was analysed only for the combination of eight Avr genes; *AvrLm1–AvrLm7* and *AvrLm9*. Since only six of these eight genes were polymorphic in the current study, 64 (i.e. 2^6) possible combinations of Avr alleles could be expected to occur in the collection of isolates. However, only 11 distinct races of *L. maculans* were identified (Table 3). This small number of races is, however, consistent with the fact that, for three Avr loci (i.e. AvrLm3, AvrLm6 and AvrLm7), one of the two alleles was at a very low frequency. Considering only the three loci with a frequency >5% for each allele (i.e. *AvrLm1*, *AvrLm4* and *AvrLm5*), all possible combinations of the avirulent and virulent alleles were identified in the collection. Only two races, Av5-6-7-(8) and Av1-5-6-7-(8), were found at all 20 sites sampled, and these two races represented more than 75% of the whole population (Table 3). In contrast, four races were each at a frequency <1%.

Table 3. Leptosphaeria maculans races identified in the large-scale French survey of experimental OSR sites

Race[a]	Frequency (%)	No. sites[b]
Av5-6-7-(8)	64.52	20
Av1-5-6-7-(8)	13.20	20
Av6-7-(8)	8.11	16
Avl-6-7-(8)	5.20	16
Av4-5-6-7-(8)	5.15	18
Av4-6-7-(8)	2.52	13
Av1-4-5-6-7-(8)	0.56	9
Avl-4-6-7-(8)	0.50	5
Avl-3-5-6-7-(8)	0.11	2
Av3-5-6-(8)	0.05	1
Av5-7-(8)	0.05	1

[a]Race nomenclature according to Balesdent et al. (2005): the figures indicate the Avr loci for which the isolate is avirulent. Figures between brackets indicate that the corresponding Avr locus has not been characterised.
[b]Number of sites where the race was found.

The number of races identified at each individual site ranged from 3 to 8 (Table 2). No significant differences between regions were observed for the richness in races, as estimated by the Margalef index (F=1.716, p=0.196) or the Simpson diversity index (F=0.09, p=0.993), or for the number of virulences per isolate (F=2.444, p=0.086). In contrast, sites from the same region showed both simple and complex race structures (Table 2, Figure 3). The lowest richness (Margalef Index) and the greatest number of virulences per isolate were both detected at the three sites in the south of France, where there was a low frequency of *AvrLm4* isolates.

Race structure and disease severity

The multi-site assessment of *B. napus* cultivars or lines for resistance to stem canker included a range of *B. napus* cvs differing in their R gene composition. There was a correlation between the good resistance of the *Rlm7* line (always ranked as "least diseased" or even immune) and the absence of isolates virulent for this R gene in the population. To correlate the frequencies of *AvrLm1* and *AvrLm4* and the stem canker severity at harvest, but take into account differences between sites in the severity of epidemics, the mean G2 ratings were first converted into a "relative G2 index" for each site and each cv. This relative index was calculated by normalising the G2 rating against that of the most susceptible cv. (i.e. Bristol). Bristol was also chosen as the reference cultivar because it did not contain resistance genes that could recognise any isolate in the *L. maculans* populations. This "relative G2 index" was correlated to the frequency of *AvrLm4* in populations for one cv. with *Rlm4* (Pollen, R^2=0.523, df=13, p<0.05) but not for Jet Neuf (also with *Rlm4*, R^2=0.028, NS) (Figure 4). In contrast, no correlation could be identified between the frequency of *AvrLm1* and the severity of stem canker on *Rlm1* cvs (either Columbus or Vivol, data not shown).

Discussion

In this paper, we report data from a large-scale collection of French *L. maculans* isolates, that

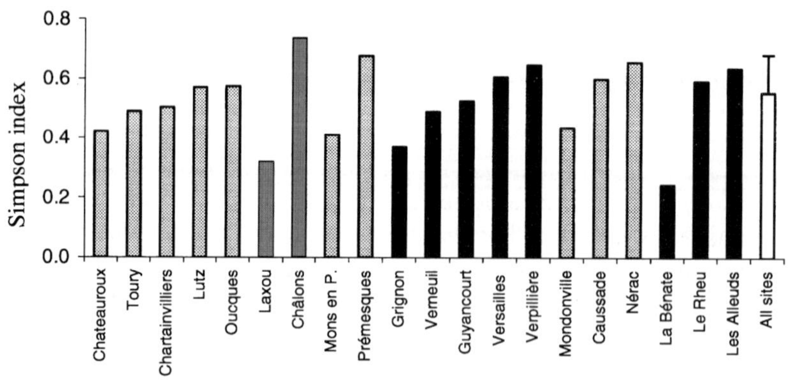

Figure 3. Values of Simpson indices of richness for isolates of *L. maculans* sampled at 20 experimental sites across France. Sites are ordered by region (alternate grey bars or black bars) as described in Table 2. White bar, mean over all sites + standard deviation.

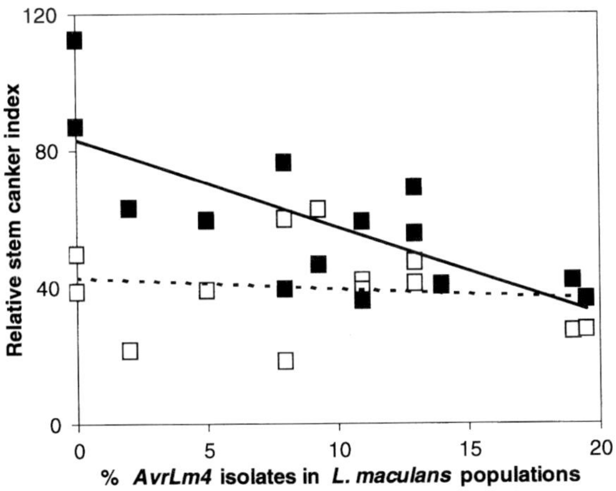

Figure 4. Relationship between the frequency (%) of *AvrLm4* isolates and the relative stem canker index on two cvs with the corresponding resistance gene *Rlm4*, Pollen (■) and Jet Neuf (□) at 14 sites across France. Regression lines: Pollen (—), Jet Neuf (- - - -). Stem canker index was evaluated using the G2 index (0, healthy, 9, severely diseased) and the data were then expressed as a percentage of the G2 index of Bristol, the most susceptible cv over all sites.

were characterised at eight Avr loci. Studies that aim to analyse the race structure or pathogenicity grouping of *L. maculans* populations often lack information on the host genotypes from which isolates originate or the R genes they harbour is often not known (Koch et al., 1991; Kutcher et al., 1993; Mahuku et al., 1997; Pongam et al., 1999; Barrins et al., 2004; Balesdent et al., 2005). In the present study, a spring type cultivar (Drakkar) that was characterised for its lack of any known resistance gene was chosen as a trap cultivar on which ascospores from any race of *L. maculans* could produce a leaf lesion. Previous work had demonstrated that winter-type cvs harboured R genes more frequently than spring type cvs (Rouxel et al., 2003b). Additional screening of more than 60 French commercial winter-type OSR cvs demonstrated that they all possess at least one gene for resistance to *L. maculans* (Balesdent and Pinochet, unpublished data). Therefore, although spring cvs are never sown in autumn under French climatic conditions, only a spring-type cv sown alongside the winter-type cvs could be used to trap *L. maculans* isolates without introducing a sampling bias. Leaf disease indexes measured on both Drakkar and winter-type cvs during autumn 2000 (Table 1) indicated that Drakkar was effectively infected by *L. maculans* in autumn, in a proportion similar to that of winter-type cvs. Leaves were sampled within a few days after the first phoma leaf spots were visible, to ensure that the collection was representative of the initial release of ascospores deposited on a given crop. Molecular analyses of a subset of this collection of isolates from three sites, and of isolates originating from different lesions on the same leaves, demonstrated that nearly all these isolates possessed distinct haplotypes and therefore did not result from a secondary cycle of conidial infection (L. Gout, pers. comm.).

In the present study, 1787 isolates originating from 20 fields were genotyped at 8–9 Avr loci. Only such a large-scale analysis could allow us to identify some rare alleles or races. For instance, there was only one isolate identified that could overwhelm either of the new resistance genes *Rlm6* or *Rlm7*. This information is relevant to the future management of these resistance sources in commercial cvs, as (i) these two R sources are effective against the current French *L. maculans* populations, since nearly 100% of the population is avirulent at these two loci, but (ii) the virulent alleles are already present, albeit at a very low frequency, in natural populations, even before the resistances have been introduced commercially.

One of the objectives of this work was to investigate the relationships between the frequency of a given Avr allele in a population and the effectiveness of the corresponding R gene for control of stem canker. This objective was only partly achieved, for two reasons. Firstly, the multi-site analysis of the relationships between phoma leaf spot incidence and severity in autumn, stem canker severity at harvest, frequencies of each Avr allele in the local pathogen population and susceptibility of cultivars was confounded by many uncontrolled factors that differed from one site to another, such as sowing date, time at which phoma leaf spot was assessed, accuracy of the phoma leaf spot assessment, along with epidemics which were not severe at some sites or loss of one experimental trial. Secondly, only extreme values were obtained for the frequencies of all Avr alleles, whatever the sites analysed, with either no or little polymorphism, or polymorphism within a limited range of values of Avr frequencies, such as 0–19% of *AvrLm4* isolates, or 2.7–39% of *AvrLm1* isolates. Nevertheless, a positive correlation was observed between the frequency of *AvrLm4* isolates and the severity of stem canker on cv. Pollen, possessing *Rlm4*. There may have been no such correlation for cv. Jet Neuf (also possessing *Rlm4*) because it possesses a high level of polygenic, general resistance, which explains why the severity of stem canker was less on Jet Neuf than all other cvs except the *Rlm7* line in the field experiment. In such a genetic background, the role of a specific, major R gene facing only a small proportion of the pathogen population (maximum 20% of the isolates) may be insignificant.

All field experiments demonstrated the complete resistance of *Rlm7*, which resulted in little leaf spotting or stem canker development at all experimental sites. This result clearly correlates with the very high frequency of *AvrLm7* isolates, and demonstrates, for a second pair of *AvrLm*/*Rlm* gene pairs, that a specific resistance expressed at the leaf stage is sufficient to control this monocyclic disease when the corresponding avirulence allele is prevalent in the local population. The first evidence of such a correlation was found for the gene pair *AvrLm1*/*Rlm1* (Ansan-Melayah et al., 1997), with *Rlm1* effectively controlling stem canker in France at a time when the frequency of *AvrLm1* in *L. maculans* populations was >80%. However, the commercial success of *Rlm1* rapidly provoked a shift in the population (Rouxel et al., 2003a), confirmed during the present study, with the frequency of *AvrLm1* <20% nationally in 2000 and 2001. Knowing that commercial cultivars harbouring *Rlm7* have recently been released in France (Pinochet et al., 2004) and that virulent isolates are already present at a low, but detectable, frequency, a new selection pressure will be exerted in favour of this virulent population. One can expect a shift in the frequency of *AvrLm7*, accompanied by a progressive loss in effectiveness of this resistance source, over the next few years, except if a significant fitness cost is associated with the gain of virulence to *Rlm7*, which is not yet known. Although the frequency of *avrLm1* was much higher, when *Rlm1* was released on a significant area (4.9% isolates *avrLm1* in 1994, Rouxel et al., 2003a), than that observed now for *avrLm7* isolates (<0.1% isolates *avrLm7* in 2000–2001), the *Rlm7* gene will have to be managed carefully through information to advisory bodies and breeding companies, surveys of virulence in *L. maculans* populations and close monitoring of the resistance of *Rlm7*-cultivars. Even though the appropriate durable resistance management strategy is not yet known, advisory bodies like CETIOM have started to recommend a diversification in use of the specific resistance genes (Pinochet et al., 2004; http://www.cetiom.fr/CTMSite/page/technique/sommairef.htm).

Taking into account the eight Avr loci for which complete information was available, only 11 distinct races of *L. maculans* were identified in the large-scale survey. This relatively low diversity is explained by the fact that only three Avr alleles, *AvrLm1*, *AvrLm4* and *AvrLm5*, displayed a significant level of polymorphism. That all possible combinations of alleles at these three loci were identified in the collection supports the conclusion that the three loci are genetically independent (Balesdent et al., 2002) and the annual occurrence of genetic recombination in the life cycle of *L. maculans*. For example, the frequency of isolates with *AvrLm1* or *AvrLm4* in the whole population was 19.5 or 8.5%, respectively. It is noteworthy that the observed frequency of races combining these two Avr alleles (i.e. races Av1-4-5-6-7-(8) plus race Av1-4-6-7-(8), 1.06% in total, Table 3) is not very different from the calculated probability of finding these two Avr alleles in the same isolate (i.e. frequency of *AvrLm1* multiplied by frequency of *AvrLm4* = 1.65%). Similarly, the observed frequency of races combining *AvrLm4* and *AvrLm5* (5.71%, Table 3) is close to the calculated probability of finding them in the same isolate (7.21%). This illustrates the importance of recombination in this fungus and questions the efficiency of R-gene pyramiding as a breeding strategy.

In spite of the limited diversity in *L. maculans* races at the national scale, this large-scale survey revealed considerable variability between neighbouring sites for different parameters assessed (i.e. disease severity, frequency of a given Avr allele, or richness in races). For example, the two sites in the North region which were only 35 km apart differed greatly in the richness in races; the Margalef index value was high at Prémesques (1.53) and differed significantly from that at Mons-en Pévèle (0.8). The site with the lowest richness, La Benate (Margalef index = 0.43, only three races identified) was in a region of intensive OSR cultivation. Similarly, the large differences in *AvrLm1* frequencies between sites (i.e. from 2.7 to 39% isolates avirulent) may illustrate different stages in the evolution of *L. maculans* population, linked to differences in selection pressure from site to site. For example, the site with the highest frequency of *AvrLm1* (Le Rheu) is located in a region where OSR is not a major crop, whereas sites with very low frequencies of *AvrLm1* are often in regions of intensive OSR cropping (Grignon, Martincourt, La Benate). Finally, the race which was most frequent, Av5-6-7-(8) (64.5% of the population) is representative of what would be expected following selection by the sequential use of R genes *Rlm2*, *Rlm4* and *Rlm1* in France (Rouxel et al., 2003b), along with the very frequent occurrence of *Rlm9* in winter OSR cvs. (Balesdent et al., 2005; Balesdent and Pinochet, unpublished data). All these data suggest that, for Avr loci, the structure of populations of *L. maculans* may be determined by the local cropping history of OSR with different R genes, in spite of the lack of population structure revealed by neutral molecular markers (L. Gout, pers. comm.). Analyses of *L. maculans* populations with AFLP markers showed not only a high level of genetic diversity within a field but also significant, although lower, differences in isolate diversity between fields, both in Canada (Mahuku et al., 1997), and Australia (Barrins et al., 2004). Our study revealed that a significant proportion of the population was virulent at the AvrLm5 and AvrLm8 loci, although *Rlm5* and *Rlm8* have never been used in commercial OSR crops. This suggests that these R sources may be present in some *B. rapa*, *B. juncea* or other mustard cultivars, used on a small scale as green manure or vegetable crops, which could have selected for the virulent alleles. Also, the more ancient agricultural use of *B. rapa* and *B. juncea*, compared to the recent history of OSR cropping, may account for the presence of these virulence alleles in the population present now on OSR.

A preliminary characterisation of *L. maculans* avirulence genes in French populations was previously done through the characterisation of the isolates in the IBCN collection (Balesdent et al., 2005). Although based on a very limited number of isolates (only 13 French isolates mostly sampled in 1990 and 1992), some of the conclusions based on this limited sample were consistent with the large-scale survey done during the current study. Both the high frequency of *AvrLm6* and *AvrLm7* (100% each in the IBCN French isolates) and the high frequencies of *avrLm2*, *avrLm3* and *avrLm9* (92% for *avrLm2*, 100% for *avrLm3* and *avrLm9* in the French IBCN isolates) were confirmed by the present study. The analysis of collections of intermediate sample size (i.e. 100–200 isolates) can allow a first description of the race structure in a given country, should the information be completely lacking. Such a survey is currently being done in additional European countries (Stachowiak et al., 2004, 2006) and in Australia (M. Barbetti, pers. comm.). In these studies, the use of the sampling procedure and host plant differentials described here will facilitate the comparison of *L. maculans* population structures in different countries or continents. In addition, starting from the very detailed analysis of the *L. maculans* French population in 2000–2001 described in the current study, it also seems feasible, with a reduced work load, to up-date the information on an annual basis by analysing around 100 isolates from 2 to 3 sites in turn. The collection of isolates gathered and characterised during such studies will be a useful resource for further population genetic analysis, or for the validation of molecular markers for a given race or virulent allele (Attard et al., 2002; L. Gout, pers comm.). Such molecular markers will greatly facilitate future population surveys by combining the use of trap cultivars with the possibility to extract DNA from individual leaf lesions, as developed for large-scale PCR analysis of the *L. maculans* mini-satellite *MinLm1* (Attard et al., 2001).

Acknowledgments

This work was funded by the French Ministry of Agriculture and Fisheries and led by AMSOL (Paris, France). I. Pauchet is thanked for her contribution to the administrative coordination of the project. The authors thank all the partners who set up experimental fields, assessed disease incidence and severity, and collected infected plants for isolation of *L. maculans*. J.P. Despeghel (Monsanto SAS), M. Renard (INRA-APBV) and D. Louchart (Momont Hennette) are thanked for providing seed. L. Coudard and M. Chabirand have contributed to this work through growing plants and G. Soubigou through doing the 100s of isolations.

References

Ansan-Melayah D, Balesdent MH, Buée M and Rouxel T (1995) Genetic characterization of *AvrLm1*, the first avirulence gene of *Leptosphaeria maculans*. Phytopathology 85: 1525–1529

Ansan-Melayah D, Rouxel T, Bertrandy J, Letarnec B, Mendes-Pereira E and Balesdent MH (1997) Field efficiency of *Brassica napus* specific resistance correlates with *Leptosphaeria maculans* population structure. European Journal of Plant Pathology 103: 835–841

Attard A, Gourgues M, Gout L, Schmit J, Roux J, Narcy JP, Balesdent MH and Rouxel T (2001) Molecular characterisation and polymorphism of *MinLm1*, a minisatellite from the phytopathogenic ascomycete *Leptosphaeria maculans*. Current Genetics 40: 54–64

Attard A, Gout L, Gourgues M, Kuhn ML, Schmit J, Laroche S, Ansan-Melayah D, Billault A, Cattolico L, Balesdent MH and Rouxel T (2002) Analysis of molecular markers genetically linked to the *Leptosphaeria maculans* avirulence gene *AvrLm1* in field populations indicates a highly conserved event leading to virulence on *Rlm1* genotypes. Molecular Plant–Microbe Interactions 15: 672–682

Aubertot JN, Schott JJ, Penaud A, Brun H and Doré T (2004) Methods for sampling and assessment in relation to the spatial pattern of phoma stem canker (*Leptosphaeria maculans*) in oilseed rape. European Journal of Plant Pathology 110: 183–192

Balesdent MH, Attard A, Ansan-Melayah D, Delourme R, Renard M and Rouxel T (2001) Genetic control and host range of avirulence toward *Brassica napus* cultivars Quinta and Jet Neuf in *Leptosphaeria maculans*. Phytopathology 91: 70–76

Balesdent MH, Attard A, Kuhn ML and Rouxel T (2002) New avirulence genes in the phytopathogenic fungus *Leptosphaeria maculans*. Phytopathology 92: 1122–1133

Balesdent MH, Barbetti MJ, Li H, Sivasithamparam K, Gout L and Rouxel T (2005) Analysis of *Leptosphaeria maculans* race structure in a world-wide collection of isolates. Phytopathology 95: 1061–1071

Barrins JM, Ades PK, Salisbury PA and Howlett BJ (2004) Genetic diversity of Australian isolates of *Leptosphaeria maculans*, the fungus that causes blackleg of canola (*Brassica napus*). Australasian Plant Pathology 33: 529–536

Brun H, Levivier S, Somda I, Ruer D, Renard M and Chèvre AM (2000) A field method for evaluating the potential durability of new resistance sources: Application to the *Leptosphaeria maculans/Brassica napus* pathosystem. Phytopathology 90: 961–966

Delourme R, Pilet-Nayel ML, Archipiano M, Horvais R, Tanguy X, Rouxel T, Brun H, Renard M and Balesdent MH (2004) A cluster of major specific resistance genes to *Leptosphaeria maculans* in *Brassica napus*. Phytopathology 94: 578–583

Koch E, Song K, Osborn TC and Williams PH (1991) Relationship between pathogenicity and phylogeny based on restriction fragment length polymorphism in *Leptosphaeria maculans*. Molecular Plant–Microbe Interactions 4: 331–349

Kutcher HR, Vandenberg CGJ and Rimmer SR (1993) Variation in pathogenicity of *Leptosphaeria maculans* on *Brassica* spp based on cotyledon and stem reactions. Canadian Journal of Plant Pathology 15: 253–258

Li CX and Cowling WA (2003) Identification of a single dominant allele for resistance to blackleg in *Brassica napus* 'Surpass 400'. Plant Breeding 122: 485–488

Li H, Sivasithamparam K and Barbetti MJ (2003) Breakdown of a *Brassica rapa* subsp *sylvestris* single dominant blackleg resistance gene in *B. napus* rapeseed by *Leptosphaeria maculans* field isolates in Australia. Plant Disease 87: 752

Mahuku GS, Goodwin PH, Hall R and Hsiang T (1997) Variability in the highly virulent type of *Leptosphaeria maculans* within and between oilseed rape fields. Canadian Journal of Botany 75: 1485–1492

McDonald BA and Linde C (2002) Pathogen population genetics, evolutionary potential, and durable resistance. Annual Review of Phytopathology 40: 349–379

Pierre JG and Regnault Y (1982) Contribution à la mise au point d'une méthode de plein champ destinée à mesurer la sensibilité des variétés de colza au Phoma. Informations Techniques du CETIOM 81: 3–18

Pinochet X, Balesdent MH, Salvi F, Mestries E and Rouxel T (2004) Identification of specific plant resistance factors to Phoma (*Leptosphaeria maculans*) among winter oilseed rape varieties: interest for variety testing and for the promotion of a first step of a durable management of resistances. International Organisation for Biological and Integrated Control Bulletin 27: 101–108

Pinochet X, Mestries E, Penaud A, Delourme R, Chèvre AM, Renard M, Brun H, Bousset L, Balesdent MH, Rouxel T and Aubertot JN (2003) Towards a durable management of genetic resistances to *Leptosphaeria maculans*. Oléagineux Corps Gras Lipides 10: 208–211

Pinon J and Frey P (1997) Structure of the *Melampsora larici-populina* populations on wild and cultivated poplar. European Journal of Plant Pathology 103: 159–173

Pongam P, Osborn TC and Williams PH (1999) Assessment of genetic variation among *Leptosphaeria maculans* isolates using pathogenicity data and AFLP analysis. Plant Disease 83: 149–154

Rouxel T, Penaud A, Pinochet X, Brun H, Gout L, Delourme R, Schmit J and Balesdent MH (2003a) A 10-year survey of populations of *Leptosphaeria maculans* in France indicates a rapid adaptation towards the *Rlm1* resistance gene of oilseed rape. European Journal of Plant Pathology 109: 871–881

Rouxel T, Willner E, Coudard L and Balesdent MH (2003b) Screening and identification of resistance to *Leptosphaeria maculans* (stem canker) in *Brassica napus* accessions. Euphytica 133: 219–231

Sprague SJ, Balesdent MH, Brun H, Hayden HL, Marcroft SJ, Pinochet X, Rouxel T and Howlett BJ (2006) Major gene resistance in *Brassica napus* (oilseed rape) is overcome by changes in virulence of populations of *Leptosphaeria maculans* in France and Australia. European Journal of Plant Pathology 114: 33–40

Stachowiak A, Olechnowicz J, Jedryczka M, Rouxel T, Balesdent MH, Happstadius I, Gladders P and Evans N (2004) Frequency of avirulence alleles in field populations of *Leptosphaeria maculans* in Germany, UK and Poland.

International Organisation for Biological and Integrated Control Bulletin 27: 91–93

Stachowiak A, Olechnowicz J, Jedryczka M, Rouxel T, Balesdent MH, Happstadius I, Gladders P, Latunde-Dada A and Evans N (2006) Frequency of avirulence genes in field populations of *Leptosphaeria maculans* in Europe. European Journal of Plant Pathology : 114: 67–75

West JS, Balesdent MH, Rouxel T, Narcy JP, Huang YJ, Roux J, Steed JM, Fitt BDL and Schmit J (2002) Colonization of winter oilseed rape tissues by A/Tox^{+} and B/Tox^{0} *Leptosphaeria maculans* (phoma stem canker) in France and England. Plant Pathology 51: 311–321

West JS, Kharbanda PD, Barbetti MJ and Fitt BDL (2001) Epidemiology and management of *Leptosphaeria maculans* (phoma stem canker) on oilseed rape in Australia, Canada and Europe. Plant Pathology 50: 10–27

Yu F, Lydiate D and Rimmer RS (2004) Identification and mapping of a third blackleg resistance locus in *Brassica napus* derived from *B. rapa* subsp, *sylvestris*. Plant, Animal & Microbe Genomes XII Conference, San Diego, California, USA

European Journal of Plant Pathology (2006) 114: 67–75
DOI 10.1007/s10658-005-2931-z

Frequency of avirulence alleles in field populations of *Leptosphaeria maculans* in Europe

Anna Stachowiak[1], Julia Olechnowicz[1], Malgorzata Jedryczka[1,*], Thierry Rouxel[2], Marie-Hélène Balesdent[2], Ingrid Happstadius[3], Peter Gladders[4], Akinwunmi Latunde-Dada[5] and Neal Evans[5]
[1]*Institute of Plant Genetics, Polish Academy of Sciences, Strzeszynska 34, PL-60479, Poznań, Poland;* [2]*Unité PMDV, Institut National de la Recherche Agronomique, Route de Saint Cyr, F-78026, Versailles, France;* [3]*Svalöf Weibull AB, SE-268 81, Svalöv, Sweden;* [4]*ADAS Boxworth, Cambridge, CB3 8NN, UK;* [5]*Rothamsted Research, Harpenden, AL5 2JQ, Hertfordshire, UK;* **Author for correspondence (Phone: +48616550248; Fax: +48616550301; E-mail: mjed@igr.poznan.pl)*

Accepted 6 September 2005

Key words: blackleg, *Phoma lingam*, race structure, resistance genes, stem canker, winter oilseed rape

Abstract

This paper describes the first large-scale Europe-wide survey of avirulence alleles and races of *Leptosphaeria maculans*. Isolates were collected from the spring rape cultivar Drakkar, with no known genes for resistance against *L. maculans*, at six experimental sites across the main oilseed rape growing regions of Europe, including the UK, Germany, Sweden and Poland. Additionally in Poland isolates were collected from cv. Darmor, which has resistance gene, *Rlm9*. In total, 603 isolates were collected during autumn in 2002 (287 isolates from Germany and the UK) and 2003 (316 isolates from Poland and Sweden). The identity of alleles at eight avirulence loci was determined for these isolates. No isolates had the virulence allele *avrLm6* and three virulence alleles (*avrLm2, avrLm3* and *avrLm9*) were present in all isolates. The isolates were polymorphic for *AvrLm1, AvrLm4, AvrLm5* and *AvrLm7* alleles, with virulence alleles at AvrLm1 and AvrLm4 loci and avirulence alleles at AvrLm7 and AvrLm5 loci predominant in populations. Virulent *avrLm7* isolates were found at only one site in Sweden. Approximately 90% of all isolates belonged to one of two races (combinations of avirulence alleles), Av5-6-7 (77% of isolates) or Av6-7 (12%). Eight races were identified, with four races at frequencies less than 1%. The study suggested that *Rlm6* and *Rlm7* are still effective sources of resistance against *L. maculans* in oilseed rape in Europe. The results are comparable to those of a similar survey done in France in autumn 2000 and 2001.

Introduction

Leptosphaeria maculans (Desm.) Ces. et de Not. is an important pathogen of oilseed rape world-wide and causes considerable losses in Australia, north America and western Europe (West et al., 2001; Fitt et al., 2006). Novel sources of resistance to *L. maculans* (phoma stem canker, blackleg), harbouring race-specific resistance genes initially proved to be very effective when introduced into Europe during the 1990s (Rouxel et al., 2003a). However, it became clear that *L. maculans* populations could change under selection, particularly when there was large-scale cropping of cultivars with resistance genes (Howlett, 2004). For example, in France, cultivars with the resistance gene *Rlm1* quickly became popular with growers and were grown over a large proportion of the oilseed rape area. This resulted in rapid adaptation of the *L. maculans* population and breakdown of *Rlm1* resistance (Ansan-Melayah et al., 1997; Rouxel et al., 2003a). Similarly in Australia, breakdown of a single dominant resistance gene derived from *Brassica rapa* subsp. *sylvestris* was observed after only 2–3 years of widespread commercial cultivation (Li et al., 2003; Sprague et al., 2006).

In Europe, the life cycle of *L. maculans* begins each season when ascospores land on oilseed rape cotyledons and/or leaves in autumn. At present, both polygenic (quantitative trait loci based) resistance to *L. maculans* (Pilet et al., 1998) and at least nine specific resistance genes (*Rlm1–Rlm9*) are known in oilseed rape. *Rlm1-Rlm4, Rlm7* and *Rlm9* are from *Brassica napus* (Ansan-Melayah et al., 1998; Balesdent et al., 2001, 2002; Delourme et al., 2004), *Rlm5* and *Rlm6* from *B. juncea* (Chèvre et al., 1997; Barret et al., 1998; Balesdent et al., 2002, 2005) and *Rlm8* from *B. rapa* (Balesdent et al., 2002). These resistance genes correspond to *L. maculans* avirulence genes *AvrLm1-AvrLm9*. In this paper, we describe the structure of races of field populations of *L. maculans* across the main oilseed rape growing regions of Europe, including the UK, Poland, Germany and Sweden, by comparison to that described in a large-scale survey in France (Balesdent et al., 2006).

Materials and methods

The isolates characterised in this study originated from field experiments established in autumn 2002 at two sites in the UK 75 km apart (Boxworth, Cambridgeshire and Rothamsted, Hertfordshire) and one site in Germany (Teendorf, Lower Saxony), and field experiments established in autumn 2003 at two sites in Poland 450 km apart (Poznan, Wielkopolska and Pulawy, Lublin Region) and one site in Sweden (Svalöv, Skøne). The field plots were sown with the spring cultivar Drakkar as a "trap cultivar". Since this cultivar has no specific resistance genes against *L. maculans*, all isolates in the pathogen population are able to infect plants and cause phoma leaf lesions. One hundred leaves with phoma lesions were collected randomly at each site (Figure 1). An additional trap cultivar, the winter cultivar Darmor (with *Rlm9*), was used in Poland, as plants of cv. Drakkar were killed by the extremely low winter temperatures in this country.

Infected leaves with typical phoma spots were sent to INRA, Versailles or Institute of Plant Genetics, Polish Academy of Sciences, Poznan, where *L. maculans* was cultured from pycnidia on the lesions. Small fragments of leaves with lesions were placed in Petri dishes containing wetted tissue paper and incubated for 2–3 days to induce pycnidia to exude conidia. The conidia produced from these pycnidia were transferred to antibiotic-amended Campbells V8 agar medium (100 mg l^{-1} streptomycin sulphate and 50 mg l^{-1} ampicillin) using a sterile needle. After two or three subcultures, hyphal tips that were free of bacterial and fungal contaminants were transferred to V8 agar medium and kept at ~22 °C for 12–14 days under alternating 12 h white/12 h near-UV light. Conidia of individual isolates of *L. maculans* were suspended in sterile double-distilled water; the suspensions were adjusted to a concentration of 10 conidia ml^{-1} and stored at −20 °C for further analysis.

Leptosphaeria maculans races were identified by inoculating each isolate onto cotyledons of a set of nine differential cultivars/lines (Balesdent et al., 2005), which incorporates eight of the nine known resistance genes (Table 1). In addition to this differential set, cv. Darmor-MX, with the *Rlm6* resistance gene but not characterized at the *Rlm9* locus, was used (Table 1).

Two-week old seedlings of the differential set of cultivars/lines were inoculated with *L. maculans* isolates at the cotyledon stage. The plants were grown in plastic trays in an air conditioned glasshouse maintained with alternating 12 h periods of

Figure 1. Location of oilseed rape experimental sites sampled throughout Europe (1, Rothamsted, UK; 2, Boxworth, UK; 3, Teendorf, Germany; 4, Svalöv, Sweden; 5, Poznan, Poland; 6, Pulawy, Poland).

Table 1. List of the oilseed rape cultivars/lines used in a differential set to differentiate isolates of *Leptosphaeria maculans* in seedling pathogenicity tests

Cultivar	Resistance Gene
Columbus	*Rlm1* + *Rlm3*
Bristol	*Rlm2* + *Rlm9*
22-1-1	*Rlm3*
Jet Neuf	*Rlm4*
150-2-1	*Rlm5*
Darmor MX[a]	*Rlm6*
23-1-1	*Rlm7*
Goéland	*Rlm9*

[a]Not characterized at the *Rlm9* locus.

20–22 °C light/16–18 °C dark. Droplets (10 μl, 10^7 conidia ml^{-1}) of spore suspensions of each of four isolates were placed on half of a cotyledon (i.e. two isolates per cotyledon, two cotyledons per seedling) that had been wounded by puncturing with a sterile needle before inoculation. Each isolate was screened on 12 plants of every differential cultivar/line, with four isolates per seedling. Trays with inoculated plants were covered with a plastic propagator lid, kept in darkness for 48 h and then transferred to a growth chamber with alternating 12 h periods at 24 °C (light)/16 °C (dark) at 70% relative humidity. Host response was scored using a 0–6 scale, where a 1–3 score (small to medium sized necrotic spots) was recorded as a resistant reaction (i.e. the isolate was avirulent on the cultivar) and a 4–6 score (grey-green tissue collapse with or without production of pycnidia) as a susceptible reaction (i.e. the isolate was virulent) (Balesdent et al., 2001).

In total, 603 isolates were studied; 207 isolates from Poland, including 147 from west Poland (Poznan collected in 2003) and 60 from east Poland (Pulawy in 2003), 203 from the UK including 103 from East Anglia (Boxworth in 2002) and 100 from the south east (Rothamsted in 2002), 84 from Germany (Teendorf in 2002) and 109 isolates from Sweden (Svalöv in 2003). Most isolates from Poland (88.9%) were from cv. Darmor (184 isolates; 135 from west and 49 from east Poland) although 23 isolates were from cv. Drakkar (12 from west and 11 from east Poland). The data were analysed by analysis of variance. In addition, two indices were used to analyse diversity of populations. These were the Margalef index, which measures the richness in species (in this case, races) of a population, and the Simpson index of diversity, which also takes into account the evenness of races within each population (Magurran, 1988; Balesdent et al., 2006).

Results

Frequencies of avirulence alleles on a European scale

All isolates were virulent on cv. Bristol (*Rlm2*), line 22-1-1 (*Rlm3*) and cv. Goéland (*Rlm9*). Thus, the avirulence alleles *AvrLm2, AvrLm3* or *AvrLm9* were absent from the 603 isolates tested (Figure 2). In contrast, all isolates were avirulent (possessed the *AvrLm6* allele) on Darmor-MX (*Rlm6*). Similarly, 99.5% of isolates possessed the *AvrLm7* allele, with only three isolates from the Swedish site virulent at the AvrLm7 locus. Three avirulence loci (AvrLm1, AvrLm4 and AvrLm5) were polymorphic, with most isolates avirulent at the AvrLm5 locus (86%), 8.4% of isolates avirulent at the AvrLm1 locus and 2.3% of isolates avirulent at the AvrLm4 locus.

Differences between countries for individual avirulence alleles

Frequencies of alleles *avrLm2, avrLm3, AvrLm6* and *avrLm9*, which were fixed in the *L. maculans* population, did not differ between the different countries. The remaining avirulence alleles studied were polymorphic. Frequencies of the avirulence alleles ranged from 1.2% (Teendorf, Germany) to 17.6% (Rothamsted, UK) for *AvrLm1*, 0% (Svalöv, Sweden and Poznan, Poland) to 6.8%

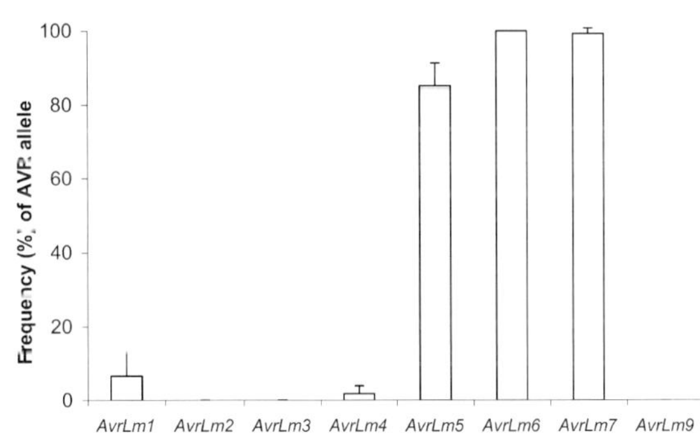

Figure 2. Mean frequency (%) of avirulence alleles in isolates from *Leptosphaeria maculans* populations sampled by isolating from phoma leaf spots on crops in the UK, Germany, Sweden or Poland. Error bars show standard deviation of AVR allele frequencies between experiment sites.

(Rothamsted, UK) for *AvrLm4* and 76.9% (Svalöv, Sweden) to 91.5% (Pulawy, Poland) for *AvrLm5* (Figure 3). Two avirulence loci, AvrLm1 and AvrLm5, were polymorphic at all experiment sites, with significant differences between sites for both alleles (P = 0.008 and P = 0.026, respectively). Amongst isolates at four sites, the AvrLm4 locus was polymorphic, but the *AvrLm4* allele was absent in isolates from west Poland and Sweden. *AvrLm7* was polymorphic at only one site (Svalöv, Sweden) where *avrLm7* isolates were observed at a low frequency (three isolates).

There were differences between the populations from the two sites sampled in the UK, with approximately twice the frequency of avirulence alleles *AvrLm1* and *AvrLm4* at Rothamsted, Hertfordshire as at Boxworth, Cambridgeshire (Figure 3). The populations of *L. maculans* from the UK and from east Poland contained more avirulence alleles than those from other sites, especially for *AvrLm1* and *AvrLm4*. The proportions of *AvrLm1* and *AvrLm4* alleles in German isolates were less than for the UK sites. In Poland, there were fewer isolates with *AvrLm1* or *AvrLm4* than at UK sites. The mean frequency of the *AvrLm1* allele for two sites in Poland was 9.7%, whereas for the two UK sites it was 14%. For *AvrLm4*, the frequencies were 1.5% for Poland and 4.8% for the UK. When the two sampling sites from Poland were compared, Pulawy (east) had more isolates (12%) with the avirulence allele *AvrLm1* than Poznan (west) (7%). At Poznan, no isolates with the *AvrLm4* allele were found. The frequency of isolates with *AvrLm1* and *AvrLm4* alleles at Pulawy resembled that at Boxworth, UK, with a slightly higher frequency of isolates with *AvrLm1* (12.4% at Pulawy, 10.4% at Boxworth) and an identical frequency of *AvrLm4* (2.9% at each site). The proportion of isolates containing the *AvrLm5* allele ranged from 76.9% in Sweden to 89.7% in Poland. The mean number of avirulence alleles differed between sites (P = 0.015). The lowest frequency was at Svalöv, Sweden (2.8%) and the highest at Pulawy, Poland (3.1%).

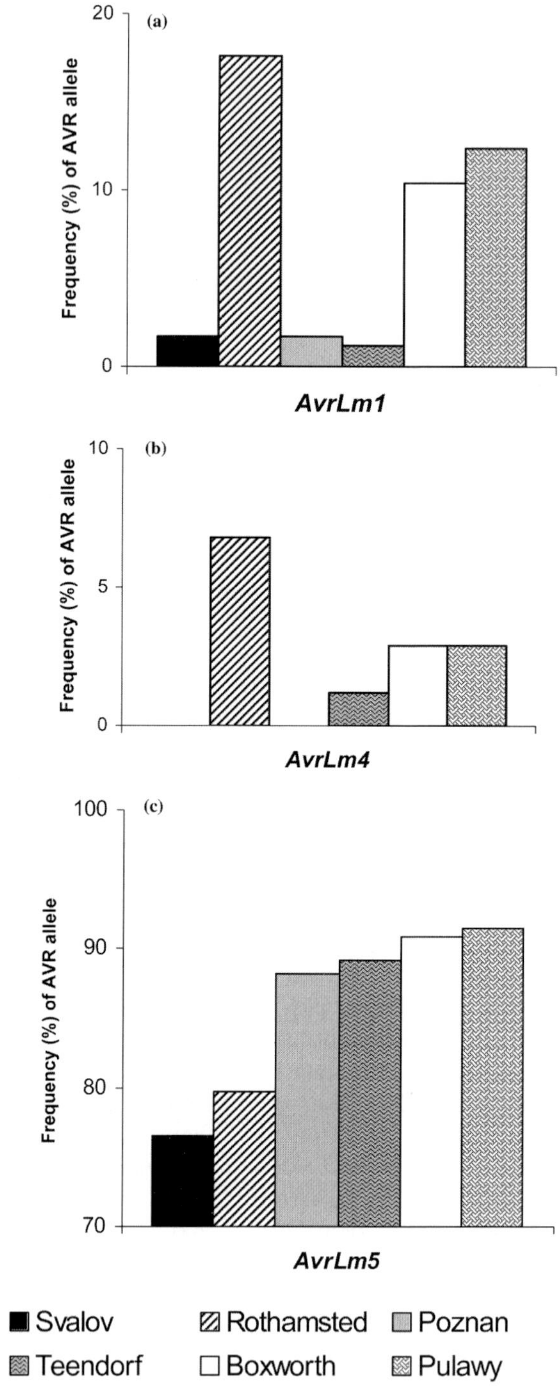

Figure 3. Differences in frequency (%) of avirulence allelles *AvrLm1* (a), *AvrLm4* (b) and *AvrLm5* (c) in field populations of *Leptosphaeria maculans* isolated from phoma leaf spots on leaves of oilseed rape at six sites across Europe.

Race structure on a European scale

Theoretically, nine avirulence genes can produce 512 races, defined as different combinations of virulence/avirulence alleles. However, only eight allele combinations were found in our survey. The most frequent race at all locations (465 isolates out of 603), with the combination of alleles *avrLm1-avrLm2-avrLm3-avrLm4-AvrLm5-AvrLm6-AvrLm7*

-*avrLm9*, was race Av5-6-7-(8) using the terminology proposed by Balesdent et al. (2005). The next most frequent race was Av6-7-(8) (78 isolates). Altogether, the two most frequent races accounted for 90% of all isolates and the other six races had very low frequencies. There were 39 isolates of race Av1-5-6-7-(8), with *AvrLm1*, *AvrLm5*, *AvrLm6* and *AvrLm7* avirulence alleles. For the three most frequent races, there were differences between sites (P = 0.003, 0.044 and 0.047). There were eight isolates of race Av4-5-6-7-(8), five isolates of race Av1-6-7-(8) and three isolates each of races Av4-6-7-(8) and Av6-(8) (only *AvrLm6* allele found). There was only one isolate of the rarest race (frequency 0.2%) with the highest number of avirulence alleles, Av1-4-5-6-7-(8). The only avirulence allele present in all isolates was *AvrLm6* (Figure 2).

Difference in race structure between countries

The race Av5-6-7-(8) was predominant at all sites, with a frequency which ranged from 60% of the population at Rothamsted (UK) to 87% at Teendorf (Germany). The frequency of Av6–7-(8), the second most abundant race, varied from 7% at Pulawy (Poland) to 19% at Svalöv (Sweden). Race Av1-5-6-7-(8) was most frequent at Rothamsted, UK (15%) and present in Germany (1.2%) and Sweden (0.9%) at low frequencies. Other races were present at low frequencies, from 3.9% to 0.5%. Overall, differences between countries for major races were not as important as those for minor races, which were rare or absent at some sites but abundant at others [e.g. race Av1-5-6-7-(8)] (Table 2).

Only the three most frequent races [Av5-6-7-(8), Av6-7-(8), Av1-5-6-7-(8)] were observed at all sites. Races Av4-5-6-7-(8) and Av1-6-7-(8) were both found at four sites, including Boxworth and Rothamsted (UK). Race Av4-5-6-7-(8) was found in Germany and east Poland, but not west Poland or Sweden. In contrast, race Av1-6-7-(8) was present in west Poland and Sweden but not Germany or east Poland. Race Av4-6-7-(8) was found only at Rothamsted and Pulawy. The rarest race, with the highest number of avirulence alleles [Av1-4-5-6-7-(8)], was found only at Rothamsted (UK) (one isolate). The race with the smallest number of avirulence alleles [Av6-(8)] occurred only in Sweden (three isolates).

The most diverse population of isolates of *L. maculans* was at Rothamsted, with seven races including the unique race Av1-4-5-6-7-(8). Consequently, both the Margalef index and the Simpson index of diversity had the highest values at this site (Table 2). There were five races at Boxworth, Svalöv and Pulawy. Six different races were found in Poland; the three most frequent races were at both sites whilst the three rarer ones were present at one site each. The populations of *L. maculans* from west Poland and Germany were least diverse, with only four races.

Table 2. The frequency of *Leptosphaeria maculans* races on oilseed rape at different sites in the UK, Germany (D), Sweden (S) or Poland (PL)

Race [a]	Frequency (%)						
	Boxworth (UK)	Rothamsted (UK)	Teendorf (D)	Svalöv (S)	Poznan (PL)	Pulawy (PL)	Mean
Av5-6-7-(8)	77.9	60.2	86.8	76.0	81.4	80.0	77.1
Av6–7-(8)	8.8	16.3	10.8	19.3	11.6	6.6	12.2
Av1-5-6-7-(8)	9.4	14.7	1.2	0.9	6.5	10.0	7.1
Av4-5-6-7-(8)	2.9	3.9	1.2	0.0	0.0	1.7	1.6
Av1-6-7-(8)	1.0	2.0	0.0	0.9	0.5	0.0	0.7
Av4-6-7-(8)	0.0	2.0	0.0	0.0	0.0	1.7	0.6
Av6-(8)	0.0	0.0	0.0	2.9	0.0	0.0	0.5
Av1-4-5-6-7-(8)	0.0	0.9	0.0	0.0	0.0	0.0	0.2
Number of isolates	103	100	84	109	147	60	100
Number of races	5	7	4	5	4	5	5
Margalef index	0.86	1.30	0.68	0.85	0.60	0.98	1.09
Simpson diversity index	0.38	0.60	0.24	0.39	0.32	0.35	0.39

[a]Race nomenclature according to Balesdent et al. (2005). The numbers indicate the AvrLm locus for which the isolate is avirulent. The number in parentheses indicates that the corresponding locus (AvrLm8) has not been characterized.

Discussion

This study reveals similarities across Europe in the frequency of avirulence alleles and structure of *L. maculans* races. The results of this European survey are comparable to those of a similar survey in France (Balesdent et al., 2006). Both surveys used the same methodology and a similar differential set of oilseed rape genotypes. The French survey was done on *L. maculans* populations in autumn 2000 and 2001, whereas the European survey is of field populations sampled in autumn 2002 (UK, Germany) and 2003 (Poland, Sweden). The observation that no isolates of *L. maculans* possessed the avirulence alleles *AvrLm2, AvrLm3* or *AvrLm9* is consistent with the French survey, which found that virulence alleles *avrLm2* and *avrLm9* were fixed in the *L. maculans* population and more than 99% of isolates (out of 1787) had the *avrLm3* allele. Similarly, all isolates sampled during the European survey were avirulent on Darmor-MX (with *Rlm6*), which is consistent with the French results, where only one isolate was virulent at the avrLm6 locus (Balesdent et al., 2006).

However, there were some differences between the two surveys in proportions of avirulence alleles within the populations. For example, in the European survey *AvrLm1* and *AvrLm4* avirulence alleles were at frequencies of 8% and 2%, respectively, whereas in the French survey the proportions of isolates avirulent on *Rlm1* and *Rlm4* were greater (20% and 9%, respectively). There were differences between the surveys in sampling sites and seasons. The isolates in the French survey were mostly collected in autumn 2000, and another survey in France showed that a rapid decrease in frequency of the *AvrLm1* avirulence allele occurred over three seasons of intensive cultivation of cultivars with the *Rlm1* resistance gene (Rouxel et al., 2003a). It is likely that a similar decrease could have occurred in other European countries. For example in Sweden, the two most common cultivars, Capitol and Cadillac (both with *Rlm1*, MH Balesdent and X Pinochet, France, unpubl.), accounted for 55% of the oilseed rape area in 2002 and could have selected for *avrLm1* isolates.

The contrasting popularities of cultivars with different *Rlm* resistance genes probably explain the different ratios of avirulence alleles at different sites. In France, sites with frequencies of the *AvrLm1* allele < 10% were in regions with intensive cropping of oilseed rape. In the European survey, the popularity of oilseed rape cultivars with the *Rlm1* resistance gene explains the low frequency of *AvrLm1* allele. The frequency of the *AvrLm5* allele in populations was similar in the French (84%, Balesdent et al., 2006) and European surveys (86%). To our knowledge, the *Rlm5* resistance gene has not been used in commercial cultivars of oilseed rape. The detection of virulent *avrLm5* isolates may be explained by the widespread use of *B. juncea* (source of *Rlm5*) for the production of mustard or as a green manure. On a local scale, this could have selected against *AvrLm5* and increased the frequency of virulent *avrLm5* isolates.

The results from the European and French surveys are also similar for *AvrLm7*. In the European survey, 99% of isolates were avirulent (*AvrLm7*) with only three isolates from Sweden with the *avrLm7* allele. In the French survey, there was only one *avrLm7* isolate out of 1787 isolates sampled (Balesdent et al., 2006). Selection for *avrLm7* might have resulted from extensive cultivation of swedes (*Brassica napus* var. *napobrassica*), of which some genotypes possess the *Rlm7* resistance gene. However, swedes are cultivated on only 150–200 ha in Sweden at Kulla-halvön and Östergötland, 80 and 400 km away from the experimental site at Svalöv. Thus, occurrence of *avrLm7* at Svalöv is unlikely to be associated with cultivation of swede. However, new sources of resistance, including *Rlm7*, are more frequently found in the older oilseed rape cultivars than in new breeding material (Rouxel et al, 2003b). Old cultivars, some of which may possess the *Rlm7* resistance gene, were grown in Sweden for many years before the 1990s. This might explain the occurrence of *avrLm7* isolates in the Swedish *L. maculans* population.

It is significant that in all five countries studied (France, Germany, Poland, Sweden and the UK) several races had the virulence allele *avrLm5*. Thus, isolates with this virulence allele have been detected before the introduction of the corresponding *Rlm* gene into commercial cultivars. This finding highlights the problems associated with the use of new resistance sources with specific, major resistance genes. The widespread use of resistant cultivars or the unregulated introduction of new resistance genes may rapidly result in adaptation of *L. maculans* to the increased selection and decrease the efficiency of the new resistance (Rouxel et al., 2003a; Howlett, 2004; Sprague et al., 2006). Resistance appears to be more durable if it is polygenic, as in cv. Jet Neuf (Pilet et al., 2001; Delourme et al., 2004). Another strategy for maintaining durability of resistance will be the careful, well-planned deployment of specific resistance genes in a "diversification" programme, where cultivars with known resistance genes are used in rotation (Aubertot et al., 2005).

The *L. maculans* population from Rothamsted (UK) had the greatest diversity, not only in terms of individual avirulence alleles, but also in terms of race structure. A number of oilseed rape cultivars and lines are grown annually on Rothamsted field plots for experimental purposes. Between 1999 and 2002, for example, these included cultivars Apex, Bristol, Lipton, Capitol, Madrigal, Synergy, Jet Neuf, Shannon, Regina, Falcon and Pronto. The selection on the *L. maculans* population imposed by this genetically heterogeneous host background may explain the diversity of avirulence alleles at this site. However, the European survey indicates that, despite different cropping ratios, growing conditions, oilseed rape cultivation practices and cultivars grown in each country, the race composition (for the three most common races) of the *L. maculans* population was similar in different countries.

Analysis of the populations from Poland suggests that wind dispersal of *L. maculans* ascospores and exchange of infected seed have played an important role in dissemination of the pathogen. Before the mid-1990s, a large proportion of the Polish oilseed rape area was sown with local cultivars and epidemics of phoma stem canker were not severe (Jedryczka et al., 1994, 1999a, b). During the mid-1990s, German and French cultivars were widely adopted in Poland and, after 2–3 seasons, a rapid increase in the incidence and severity of stem canker was observed. The Polish *L. maculans* race structure is currently very similar to that in western Europe, which suggests that the initial distribution of *L. maculans* might have been seed-borne. However, most seed production is done within Poland and only seed of new cultivars is likely to be imported. Thus, the Polish population might have been expected to have been the most diverse for number of avirulence genes present. Most isolates from Poland were collected from cv. Darmor, which has *Rlm9*. Therefore, only isolates with the *avrLm9* allele could infect these plants. However, of the 23 Polish isolates collected on cv. Drakkar, all had the *avrLm9* virulence allele, as in other European countries (Balesdent et al., 2005b). Therefore the Polish *L. maculans* population probably does not differ from populations in other countries with respect to the AvrLm9 locus. The increasing importance of phoma stem canker in Poland has also been connected with climate change associated with global warming (West et al., 2005). Widespread release of *L. maculans* ascospores in the autumn produces phoma leaf spots on crops before winter (West et al., 2002) and occurrence of mild winters decreases loss of infected leaves (Huang et al., 2005), thereby allowing the pathogen to colonise the stem and carry over to the next growing season. This hypothesis is examined in a detailed study of the Polish *L. maculans* and *L. biglobosa* populations (Jedryczka et al., 2004).

The little information available on resistance genes in current and historical oilseed rape cultivars (Ansan-Melayah et al., 1995, 1998; Balesdent et al., 2001, 2002; Rouxel et al., 2003b; Delourme et al., 2004, 2006) limits interpretation and understanding of changes and evolution of avirulence genes and race composition in *L. maculans* populations. The European survey suggests that the *Rlm6* and *Rlm7* resistance genes are likely to be effective sources of resistance against *L. maculans* in Europe. The example of the *Rlm1* resistance gene showed that a specific resistance operating at the cotyledon and leaf stage is sufficient to control stem canker epidemics, at least for a few seasons (Ansan-Melayah et al., 1997; Rouxel et al., 2003a; Sprague et al. 2005). Undoubtedly, such specific resistance will be of only short-term value unless combined with quantitative (polygenic) resistance or used in combination with other effective resistance genes. Numerous studies show virulence is achieved or lost through a set of small, highly conserved genetic events (Joosten et al., 1994; Bryan et al., 2000; Attard et al., 2002). Therefore, knowledge of race structure of a pathogen population before and after the introduction of a new resistance gene is essential for the good management of resistance to maintain its durability over several seasons. The recent survey of Balesdent et al. (2006) suggests that a sample of *ca.* 100–200 isolates is sufficient to draw general conclusions about frequency of avirulence genes and race structure in *L. maculans* populations at a given site. However, this sample size may be too small to detect virulence alleles present at very low frequencies, which is usually the case when a new resistance gene is introduced in an area.

Acknowledgements

This work was funded by the European Commission under Framework Programme 5, Quality of Life (SECURE project, contract number QLK5-CT-2002–01813) and the UK Department for Environment, Food and Rural Affairs (OREGIN project, ARO703) and the Biotechnology and Biological Sciences Research Council. Some experiments were done by staff of the Department of Plant Pathology, Agricultural University in Poznan, Poland and we are grateful for technical assistance during these experiments.

References

Ansan-Melayah D, Balesdent MH, Buée M and Rouxel T (1995) Genetic characterization of *AvrLm1*, the first avirulence gene of *Leptosphaeria maculans*. Phytopathology 85: 1525–1529

Ansan-Melayah D, Balesdent MH, Delourme R, Pilet ML, Renard M, Tanguy X and Rouxel T (1998) Genes for race-specific resistance against blackleg disease in *Brassica napus* L. Plant Breeding 117: 373–378

Ansan-Melayah D, Rouxel T, Bertrandy J, Letarnec B, Mendes-Pereira E and Balesdent MH (1997) Field efficiency of *Brassica napus* specific resistance correlates with *Leptosphaeria maculans* population structure. European Journal of Plant Pathology 103: 835–841

Attard A, Gout L, Gourgues M, Kuhn ML, Schmit J, Laroche S, Ansan-Melayah D, Billault A, Cattolico L, Balesdent MH and Rouxel T (2002) Analysis of molecular markers genetically linked to the *Leptosphaeria maculans* avirulence gene *AvrLm1* in field populations indicates a highly conserved event leading to virulence on *Rlm1* genotypes. Molecular Plant-Microbe Interactions 15: 672–682

Aubertot JN, West JS, Bousset-Vaslin L, Salam MU, Barbetti MJ and Diggle AJ (2006) Improved resistance management for durable disease control: a case study of phoma stem canker of oilseed rape (*Brassica napus*) European Journal of Plant Pathology 114: 91–106

Balesdent MH, Attard A, Ansan-Melayah D, Delourme M and Renard M (2001) Genetic control and host range of avirulence towards *Brassica napus* cultivars Quinta and Jet Neuf in *Leptosphaeria maculans*. Phytopathology 91: 70–76

Balesdent MH, Attard A, Kühn ML and Rouxel T (2002) New avirulence genes in the phytopathogenic fungus *Leptosphaeria maculans*. Phytopathology 92: 1122–1133

Balesdent MH, Barbetti MJ, Hua Li, Sivasithamparam K, Gout L and Rouxel T (2005) Analysis of *Leptosphaeria maculans* race structure in a world-wide collection of isolates. Phytopathology 85: 1061–1071

Balesdent MH, Louvard K, Pinochet X and Rouxel T (2006) A large-scale survey of races of *Leptosphaeria maculans* occurring on oilseed rape in France. European Journal of Plant Pathology 114: 53–65

Barret P, Guérif J, Reynoird JP, Delourme R, Eber F, Renard M and Chèvre AM (1998) Selection of stable *Brassica napus-B. juncea* recombinant lines resistant to blackleg (*Leptosphaeria maculans*) 2 : A 'to and fro' strategy to localise and characterise interspecific introgressions on *B. napus* genome. Theoretical and Applied Genetics 96: 1097–1103

Bryan GT, Wu KS, Farrall L, Jai Y, Hershey HP, McAdams SA, Faulk KN, Donaldson GK, Tarchini R and Valent B (2000) A single amino acid difference distinguishes resistant and susceptible alleles of the rice blast resistance gene *Pi-ta*. Plant Cell 12: 2033–2045

Chèvre AM, Barret P, Eber F, Dupuy P, Brun H, Tanguy X and Renard M (1997) Selection of stable *Brassica napus-B. juncea* recombinant lines resistant to blackleg (*Leptosphaeria maculans*) 1: Identification of molecular markers, chromosomal and genomic origin of the introgression. Theoretical and Applied Genetics 95: 1104–1111

Delourme R, Chèvre A-M, Brun H, Rouxel T, Balesdent MH, Dias JS, Salisbury PA, Renard M and Rimmer SR (2006) Major gene and polygenic resistance to *Leptosphaeria maculans* in oilseed rape (*Brassica napus*). European Journal of Plant Pathology 114: 41–52

Delourme R, Pilet-Nayel ML, Archipiano M, Horvais R, Tanguy X, Rouxel T, Brun H, Renard M and Balesdent MH (2004) A cluster of major specific resistance genes to *Leptosphaeria maculans* in *Brassica napus*. Phytopathology 94: 578–583

Fitt BDL, Brun H, Barbetti MJ and Rimmer SR (2006) World-wide importance of phoma stem canker (*Leptosphaeria maculans* and *L. biglobosa*) on oilseed rape (*Brassica napus*). European Journal of Plant Pathology 114: 3–15

Howlett BJ (2004) Current knowledge of the interaction between *Brassica napus* and *Leptosphaeria maculans*. Canadian Journal of Plant Pathology 26: 245–252

Huang YJ, Fitt BDL, Jedryczka M, Dakowska S, West JS, Gladders P, Steed JM and Li ZQ (2005) Patterns of ascospore release in relation to phoma stem canker epidemiology in England (*Leptosphaeria maculans*) and Poland (*L. biglobosa*). European Journal of Plant Pathology 111: 263–277

Jędryczka M, Fitt BDL, Kachlicki P, Lewartowska E, Balesdent MH and Rouxel T (1999a) Comparison between Polish and United Kingdom populations of *Leptosphaeria maculans*, cause of stem canker of oilseed rape. Journal of Plant Pathology and Plant Protection 106: 608–617

Jedryczka M, Lewartowska E and Frencel I (1994) Properties of *Phoma lingam* (Tode ex Fr.) Desm. isolates from Poland.

I. Pathogenicity characterization. Phytopathologia Polonica 7: 71–79

Jedryczka M, Matysiak R and Graham K (2004) Lepto-net and SPEC - new projects to control the stem canker of oilseed rape in Poland. International Organisation for Biological Control Bulletin 27: 125–130

Jedryczka M, Rouxel T and Balesdent MH (1999b) Rep-PCR based genomic fingerprinting of a Polish population of *Leptosphaeria maculans*. European Journal of Plant Pathology 105: 813–823

Joosten MHAJ, Cozijnsen TJ and deWit PJGM (1994) Host resistance to a fungal tomato pathogen lost by a single base-pair change in an avirulence gene. Nature 367: 384–386

Li H, Barbetti MJ and Sivasithamparam K (2003) Breakdown of a *Brassica rapa* subsp. *sylvestris* single dominant blackleg resistance gene in *B. napus* rapeseed by *Leptosphaeria maculans* field isolates in Australia. Plant Disease 87: 752

Magurran AE (1988) Ecological Diversity and its Measurement, Princeton University Press, Princeton, New Jersey

Pilet ML, Delourme R, Foisset N and Renard M (1998) Identification of loci contributing to quantitative field resistance to blackleg disease, causal agent *Leptosphaeria maculans* (Desm.) ces et de Not., in winter rapeseed (*Brassica napus* L.). Theoretical and Applied Genetics 96: 23–30

Pilet ML, Duplan G, Archipiano M, Barret P, Baron C, Horvais R, Tanguy X, Lucas MO, Renard M and Delourme R (2001) Stability of QTL for field resistance to blackleg across two genetic backgrounds. Crop Science 41: 197–205

Rouxel T, Penaud A, Pinochet X, Brun H, Gout L, Delourme R, Schmit J and Balesdent MH (2003a) A ten year survey of populations of *Leptosphaeria maculans* in France indicates a rapid adaptation towards the *Rlm1* resistance gene of oilseed rape. European Journal of Plant Pathology 109: 871–881

Rouxel T, Willner E, Coudard L and Balesdent MH (2003b) Screening and identification of resistance to *Leptosphaeria maculans* in *Brassica napus* accessions. Euphytica 133: 219–231

Sprague SJ, Balesdent MH, Brun H, Hayden HL, Marcroft SJ, Pinochet X, Rouxel T and Howlett BJ (2006) Major gene resistance in *Brassica napus* (oilseed rape) is overcome by changes in virulence of populations of *Leptosphaeria maculans* in France and Australia. European Journal of Plant Pathology 114: 33–40

West JS, Fitt BDL, Jedryczka M and Karolewski Z (2005) Evidence for the eastward spread of the aggressive oilseed rape pathogen *Leptosphaeria maculans* in Europe. Plant Protection and Plant Health in Europe: Introduction and Spread of Invasive Species; BCPC Symposium Proceedings, pp. 293–294

West JS, Jędryczka M, Leech PK, Dakowska S, Huang Y-J and Fitt BDL (2002) Biology of *Leptosphaeria maculans* (stem canker) ascospore release in England and Poland. International Organisation for Biological Control Bulletin 25: 21–30

West JS, Kharbanda PD, Barbetti MJ and Fitt BDL (2001) Epidemiology and management of *Leptosphaeria maculans* (phoma stem canker) on oilseed rape in Australia, Canada and Europe. Plant Pathology 50: 10–27

European Journal of Plant Pathology (2006) 114: 77–89
DOI 10.1007/s10658-005-2643-4

Fitness cost associated with loss of the *AvrLm4* avirulence function in *Leptosphaeria maculans* (phoma stem canker of oilseed rape)

Y.-J. Huang[1,*], Z.-Q. Li[1,3], N. Evans[1], T. Rouxel[2], B.D.L. Fitt[1] and M.-H. Balesdent[2]
[1]*Rothamsted Research, Harpenden, Herts, AL5 2JQ, UK;* [2]*INRA-PMDV unit, Route de St Cyr, F-78026, Versailles Cedex, France;* [3]*Plant Protection Institute, Inner Mongolia Academy of Agricultural Sciences, 010031, Huhhot, China;* **Author for correspondence (Phone: +44-1582763133; Fax: +44-1582760981; E-mail: yong-ju.huang@bbsrc.ac.uk)*

Accepted 1 September 2005

Key words: ascospore, avirulence gene, blackleg, *Brassica napus* (canola), near-isogenic isolate, resistance gene

Abstract

Near-isogenic isolates of *Leptosphaeria maculans* differing at the AvrLm4 avirulence locus (*AvrLm4* or *avrLm4*) were produced *in vitro*. Methods for inoculation of leaves of oilseed rape with ascospores or conidia were compared. The 'ascospore shower' inoculation was the most efficient method for use when inoculum is limited (e.g. ascospores produced *in vitro*). It was used in controlled environments to compare fitness of *AvrLm4* and *avrLm4* isolates at 5, 10, 15, 20 or 25 °C on leaves of oilseed rape cultivars Eurol and Darmor lacking the resistance gene *Rlm4*, which corresponds to *AvrLm4*. At all temperatures tested, *AvrLm4* ascospores produced more lesions than *avrLm4* ascospores. The diameters of lesions produced by *AvrLm4* ascospores were greater than those of lesions produced by *avrLm4* ascospores. At 15–20 °C, more lesions initiated by *AvrLm4* ascospores produced pycnidia than did lesions initiated by *avrLm4* ascospores. However, there were no differences between *AvrLm4* and *avrLm4* isolates in incubation period (from inoculation to appearance of lesions) or rate of mycelial growth in leaves from lesions towards the stems. In field experiments with winter oilseed rape cultivars lacking *Rlm4*, the frequency of *AvrLm4* isolates increased from 5.7% at the phoma leaf lesion stage (autumn) to 20.5% at the stem canker stage (summer) during 2002/2003 and from 7.9 to 11.5% during 2003/2004 growing seasons. Results of controlled environment and field experiments indicate that *avrLm4* isolates have a fitness cost compared to *AvrLm4* isolates.

Abbreviations: AFLP – amplified fragment length polymorphism; BC – back-cross; cv – cultivar; GS – growth stage; NI – near isogenic

Introduction

Leptosphaeria maculans is an important worldwide pathogen of cruciferous crops, especially *Brassica* species, causing phoma stem base canker (blackleg) of oilseed rape in Europe, Canada and Australia (West et al., 2001; Fitt et al., 2006). Use of cultivars (cv) with resistance genes is an efficient and environmentally friendly way to control the disease. However, new major resistance genes often become ineffective after two to three seasons if cultivars carrying them are grown commercially (Li et al., 2003; Rouxel et al., 2003a) or in experimental plots (Brun et al., 2000). There is a need to evaluate *a priori* the durability of new resistance genes to design management strategies that will minimise the risk of rapid resistance breakdown.

One approach to predicting the durability of new resistance genes is through understanding the fitness associated with the corresponding pathogen avirulence genes (Leach et al., 2001). Fitness of an organism is defined as the combined ability for it to survive and reproduce (Crow, 1986; Pringle and Taylor, 2002). Fitness can be measured in absolute or relative terms. For plant pathogens, several traits, such as reproductive rate, infection efficiency or severity of disease caused, have been used to measure fitness (Vera Cruz et al., 2000; Leach et al., 2001; Pringle and Taylor, 2002). Several bacterial avirulence genes possess a function in disease symptom expression (Leach et al., 2001). Therefore, it is suggested that the fitness cost associated with pathogen evolution from avirulence to virulence to overcome host resistance affects the durability of the resistance (Vera Cruz et al., 2000). This has been demonstrated for the *Xanthomonas oryzae*/rice gene-for-gene interaction (Vera Cruz et al., 2000). However, few other experimental data sets are available to test this hypothesis, especially for fungal plant pathogens.

L. maculans has a gene-for-gene interaction with *B. napus* (Ansan-Melayah et al., 1998; Balesdent et al., 2001, 2002). Although several major genes conferring resistance to *L. maculans* have been identified (Ansan-Melayah et al., 1998; Delourme et al., 2004, 2006), it has been difficult to investigate *a priori* the durability of these resistance genes through assessing fitness associated with corresponding avirulence genes without having available pairs of *L. maculans* isolates differing only at the corresponding avirulence locus. Since pairs of near isogenic (NI) *L. maculans* isolates differing only at one avirulence locus (Attard et al., 2002) have now been developed, the potential durability of major gene resistance can be examined by comparing the fitness of corresponding avirulent and virulent isolates.

In Europe, including the UK, phoma stem canker epidemics are initiated by ascospores produced on infected debris and there is no evidence that conidia play a role in the development of epidemics (Gladders and Musa, 1980; West et al., 1999; Huang et al., 2005). Effects of major genes for resistance operating during the infection of leaves by germ tubes from these ascospores in autumn can remain for the whole growing season because *L. maculans* is a monocyclic pathogen (Fitt et al., 2006). However, most previous work on the *L. maculans*/*B. napus* interaction has been done with conidial inoculum applied to cotyledons. Only recently have experiments on leaf penetration and infection been done with ascospores (Biddulph et al., 1999; Toscano-Underwood et al., 2001; Huang et al., 2003). To obtain lesions using conidial inoculum, leaves usually have to be wounded before inoculation. Since wounding can induce plant defence responses or allow the pathogen to bypass leaf resistance, this approach is not suitable to compare the fitness of NI *L. maculans* isolates.

Since techniques were published for the *in vitro* production of ascospores of *L. maculans* by crossing the two opposite mating types (Mengistu et al., 1993; Gall et al., 1994), it has become possible to use ascospores from genetically defined isolates as inoculum to study the relative fitness of virulent and avirulent isolates. However, since only small numbers of ascospores are produced *in vitro*, methods for inoculating with the large numbers of ascospores produced on infected debris under natural conditions are not suitable for inoculating with ascospores produced from defined crosses *in vitro*.

This paper investigates the fitness cost associated with the loss of avirulence function of the *AvrLm4* allele. For this purpose, NI *L. maculans* isolates carrying either the avirulent (*AvrLm4*) or the virulent (*avrLm4*) allele of the avirulence locus AvrLm4 were compared for specific components of fitness (i.e. infection efficiency, latent period, size of the lesions, ability to sporulate on the lesions, systemic growth down the petiole) in controlled environments using a new 'ascospore shower' inoculation method. To support results from controlled environment experiments, changes in frequencies of *AvrLm4* isolates in natural *L. maculans* populations were assessed under field conditions using winter oilseed rape crops lacking the corresponding resistance gene *Rlm4*.

Materials and methods

Production of near-isogenic isolates of Leptosphaeria maculans

Near-isogenic isolates of *L. maculans*, differing in their alleles at the AvrLm4 locus, were produced by a back-crossing (BC) scheme. Two sister isolates of opposite mating-type, v23.1.2 (*avrLm4, MAT1–1*) and v23.1.3 (*AvrLm4, MAT1–2*), were crossed *in vitro* and the F_1 progeny isolates were characterised at the AvrLm4 locus by inoculation of cotyledons of cv Jet Neuf (carrying *Rlm4*) and cv Westar (susceptible with no resistance genes) (Balesdent et al., 2001; Attard et al., 2002). The mating types of progeny isolates were determined by PCR (Cozijnsen and Howlett, 2003). An *avrLm4, MAT1–1* isolate was selected at random from the F_1 progeny and back-crossed to v23.1.3 (*AvrLm4, MAT1–2*). The BC_1 progeny isolates were characterized on cv Jet Neuf and an *avrLm4, MAT1–1* isolate was selected and crossed to v23.1.3 for the second back-cross (BC_2). This procedure was repeated until the fifth BC generation (BC_5), obtained by crossing v23.1.3 and v37.1.4. Ninety-five progeny isolates recovered from the fifth BC were characterised at the AvrLm4 and *MAT* loci.

AFLP markers were used to confirm the increase in isogenicity at each step of the BC scheme. AFLP bands were amplified using 21 primer pairs (Kuhn et al., 2006) with genomic DNA of the F_1 parental isolates v23.1.2 and v23.1.3, along with the selected parental isolate for each subsequent round of BC. Bands were separated on CastAway 5.5% acrylamide gels (Stratagene) and visualized by silver nitrate staining (Kuhn et al., 2006). Bands that were polymorphic between parental isolates of each BC generation were recorded.

In vitro growth rates of NI L. maculans isolates

The *in vitro* growth rates of the F_1 parental isolates and six avirulent (*AvrLm4*) and six virulent (*avrLm4*) NI isolates from BC_5 progeny were compared. Plugs (8 mm diameter) were taken from the margin of actively growing colonies of each isolate and placed in the centre of Petri dishes (9 cm diameter) filled with 20 ml of V8 juice agar. Petri dishes were incubated in darkness at 4, 12 or 22 °C, with three or four replicate Petri dishes per isolate. Radial growth was assessed by measuring two perpendicular colony diameters after 6, 11 and 17 days (12 and 22 °C) or 10, 24 and 32 days (4 °C). To estimate the growth rate, a linear regression of colony diameter on incubation time was done for each isolate. Analyses of variance were done to assess differences between *AvrLm4* and *avrLm4* NI isolates in growth rate *in vitro* at different temperatures, using GENSTAT statistical software (Payne et al., 1993).

In vitro crosses of NI isolates to produce ascospores

To produce ascospores of isolates with either *AvrLm4* or *avrLm4* alleles, four *AvrLm4* (two *MAT1–1* and two *MAT1–2*) and four *avrLm4* (two *MAT1–1* and two *MAT1–2*) isolates from the fifth BC progeny were randomly selected for subsequent crossing. Conidial suspensions of the eight isolates were prepared and adjusted to a concentration of 10^6 spores ml^{-1}. Equal volumes of spore suspensions of the four *AvrLm4* isolates were mixed, and 50 μl of the mixture were spread over 20% V8 juice agar in Petri dishes (5 cm diameter). Petri dishes were incubated for 7 days at 25 °C, under mixed white (500 nm) and near-UV light (360 nm) for 12 h of light per day, then flooded with 5 ml of 1.5% water agar (cooled to 45 °C) and incubated for 5–6 weeks at 12 °C under alternating near-UV light (12 h per day)/darkness. Pseudothecia produced under these conditions contained ascospores of only the *AvrLm4* genotype. Similarly, pseudothecia containing *avrLm4* ascospores were produced by incubating four *avrLm4* isolates together.

Development of the 'ascospore shower' method for inoculation with ascospores produced in vitro

Seeds of cv Lipton were sown in pots (7 cm diameter) containing peat-based compost and a soluble fertiliser (1·5 kg PG mix m^{-3}; Petersfield Products, Cosby, Leicester, UK). Plants were grown in a glasshouse and thinned to one plant per pot 10 days after sowing. Pots were then placed in seed trays (37 cm×23 cm) and transferred to a 15 °C controlled-environment cabinet. Plants were inoculated when they reached growth stage (GS) 1,3 (Sylvester-Bradley and Makepeace, 1985) and kept in the growth cabinet throughout the experiment.

A suspension of *L. maculans* conidia was prepared from a 12-day-old culture of isolate L44, which was obtained from a phoma leaf spot collected during November 2001 from winter oilseed rape cv Apex at Rothamsted. A suspension of *L. maculans* ascospores was prepared from naturally infected oilseed rape stem base debris collected during August 2002 at Rothamsted, using the method described by Huang et al. (2001). Ascospores from defined crosses were obtained by crossing *L. maculans* isolates v23.1.2 and v23.1.3 on V8 juice agar (Balesdent et al., 2001). Pieces of agar with mature pseudothecia were attached to the under-side of a Petri dish lid, then the lid was placed over the base of the dish, which was then placed at 15 °C for 6–8 h in darkness to allow ascospores to discharge into the bottom of the dish. The dish was viewed under a binocular stereo-microscope (Olympus Optical Co., London) to confirm that ascospores had been discharged. Distilled water was added to suspend the ascospores. The concentration of the conidial suspension was adjusted to 10^6 conidia ml^{-1} and the concentration of the ascospore suspension was adjusted to 10^3 ascospores ml^{-1}.

For inoculation of leaves with conidia, three methods were used: (1) spray the whole plant with conidial suspension; (2) point inoculation with a 10 μl drop of conidial suspension without wounding; (3) point inoculation with a 10 μl drop of conidial suspension on a site that had been wounded using a sterile pin. Three plants were inoculated by each method. For point inoculation, only the first and second leaves of each plant were inoculated; 6–8 wounded sites or 10–12 unwounded sites on each plant were inoculated. For inoculation with ascospores, three inoculation methods were used: (1) spray the whole plant with ascospore suspension; (2) point inoculation with a 10 μl drop of ascospore suspension on the leaf surface without wounding the leaf; (3) 'rain down' dry ascospores onto the plant. To allow ascospores to 'rain down' onto plants ('ascospore shower'), three small pieces (0.5×2 cm) of oilseed rape debris with mature pseudothecia were attached to the under-side of a tray cover (37×23×14 cm) with Vaseline (Chesebrough-Pond's Ltd, London) and the debris was sprayed with distilled water until run-off. Alternatively, for ascospores produced by *in vitro* crosses, pieces of agar with mature pseudothecia produced *in vitro* were attached to the under-side of a tray cover. For both sources of inoculum, the tray cover was then placed over the tray with plants (Figure 1). Three plants were inoculated by each inoculation method. For point inoculation, 4–6 unwounded sites on each plant were inoculated. With 'ascospore shower' inoculation, the duration of the inoculation period was 2 h for ascospores from natural conditions and 24 h for ascospores from *in vitro* crosses. After inoculation, plants were covered immediately with polyethylene bags sprayed inside with distilled water to maintain leaf wetness for 72 h.

The numbers of new phoma leaf spots on each leaf of the inoculated plants were counted daily until no more new leaf spots appeared. The incubation period (from inoculation to the appearance of first phoma leaf spots) for each treatment was recorded. For point inoculation, the infection efficiency was estimated as the percentage of inoculated sites that produced lesions. To estimate the infection efficiency of conidia or ascospores inoculated by spraying, at the time of inoculation three microscope slides were placed among the plants so that conidia or ascospores used for the experiment were also deposited on the slides. The number of conidia or ascospores deposited per cm^2 on each of the slides was counted under a light microscope (Olympus Optical Co., London). The leaf areas of five additional uninoculated plants were measured at the time of inoculation. The mean number of conidia or ascospores deposited per plant was estimated [leaf area (cm^2) per plant × number of spores per cm^2]. The infection efficiency was then estimated as number of spores required to cause one lesion (i.e. total number of spores deposited per plant divided by total number of lesions which developed per plant).

Fitness of virulent (avrLm4) and avirulent (AvrLm4) L. maculans *isolates during leaf infection in controlled environments*

Plants of oilseed rape cultivars Eurol (lacking *Rlm4*) and Darmor (lacking *Rlm4*) were grown in

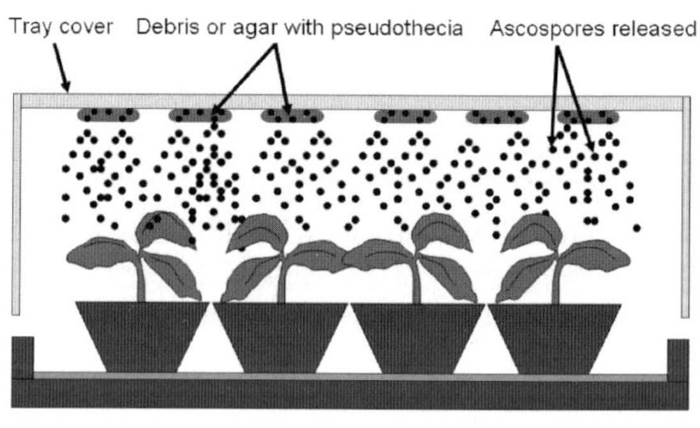

Figure 1. 'Ascospore shower' method for inoculation of oilseed rape seedlings with *Leptosphaeria maculans* ascospores produced under natural conditions on oilseed rape debris or *in vitro* by crossing the two opposite mating types on agar medium. Small pieces of oilseed rape debris with mature pseudothecia were attached to the under-side of a tray cover and the debris was sprayed with distilled water until run-off. Alternatively, for defined crosses, pieces of agar with mature pseudothecia produced *in vitro* were attached to the under-side of a tray cover. The tray cover was then placed over the plants to allow dry ascospores to 'rain down' onto the plants.

pots (5 cm diameter) containing peat-based compost and a soluble fertiliser. The pots were put in trays (37×23 cm) in four rows with seven pots in each row. The two cultivars were sown in alternate rows. Plants were grown in a glasshouse, thinned to one plant per pot 10 days after sowing and then transferred to a 15 °C controlled-environment cabinet until they reached GS 1, 3 and were ready for inoculation.

Plants were inoculated using the 'ascospore shower' method with ascospores of either *AvrLm4* or *avrLm4* isolates that had been produced *in vitro*. A total of 10 trays with 28 plants per tray was inoculated. After inoculation, plants were sprayed with distilled water and covered with tray covers. Two trays (one inoculated with ascospores of *AvrLm4* isolates, the other inoculated with ascospores of *avrLm4* isolates) were selected at random and transferred to each of the 5, 10, 15, 20 or 25 °C growth cabinets. The tray covers were kept on for 48 h to maintain leaf wetness. The experiment was repeated three times, with each experiment inoculated with fresh ascospores produced in a new set of crosses. Growth cabinets were allocated at random to each of the three experiments, whenever possible. The temperatures in the growth cabinets were monitored throughout the experiments and varied by ±1 °C.

Plants were assessed daily until no new leaf spots appeared and the incubation period was recorded. The diameters of 10 lesions on plants grown at 15, 20 and 25 °C were measured 13–14 days after inoculation. To compare the growth of *AvrLm4* and *avrLm4* isolates after initial infection (growth from leaf lesion towards the petiole to reach the stem), affected leaves were detached 18–19 days after inoculation, and *L. maculans* was isolated at 1, 2 and 3 cm from centre of the lesion in the direction of petiole, using the method described by West et al. (2002). To compare the sporulation of *AvrLm4* and *avrLm4* isolates, the number of lesions that produced pycnidia was counted at the end of each experiment (i.e. when no more new lesions appeared).

Analyses of variance using the GENSTAT statistical software compared components of the fitness (e.g. incubation period, number of lesions, size of lesions, growth *in planta* and sporulation ability) of *AvrLm4* and *avrLm4* isolates during infection of leaves of oilseed rape without the resistance gene *Rlm4* (Payne et al., 1993). Linear regressions of incubation period on temperature were calculated separately for the inoculum (*AvrLm4* or *avrLm4*) treatments in each of the three replicate experiments. Analyses of position and parallelism were done to assess whether the data were best fitted by a single line or series of parallel lines or series of non-parallel lines for *AvrLm4* and *avrLm4* isolates.

Fitness of virulent (avrLm4) and avirulent (AvrLm4) isolates in natural populations of L. maculans

Oilseed rape cultivars Darmor, Mohican and Pactol, all lacking resistance gene *Rlm4*, were grown in field experiments during two growing seasons at Versailles, France. During 2002/2003, each of the three cultivars was grown in 6.5×2.3 m plots with four replicates. During 2003/2004, only cv Mohican was grown in 20×3 m plots with 10 replicates. The experimental plots for the two seasons were located in the same large field, but not at the same place in the field. During 2002/2003, 2.87 ha of oilseed rape (cultivars Extra and Pollen, with *Rlm4*) were also grown in this field. Isolates were recovered from individual phoma leaf spot lesions sampled from plots in autumn (5 and 11 November 2002, 24–26 November 2003) or spring (9 April 2004), from stem cankers sampled before harvest during summer (20 June 2003, 8 July 2004) and from ascospores produced on the debris in the following autumn using the method described by West et al. (2002). All isolates were characterised at the AvrLm4 locus using cotyledon inoculation tests on cultivars Jet Neuf (carrying *Rlm4*) and Westar (susceptible control with no resistance genes) (Balesdent et al., 2001).

Results

Production of near-isogenic isolates of Leptosphaeria maculans

The 21 AFLP primer pairs generated 94 AFLP fragments that were polymorphic between the two parental isolates of the F_1 generation. The number of polymorphic markers decreased at each successive BC generation (Table 1), indicating an increase in isogenicity. The level of polymorphism at each BC generation was greater than expected (Table 1). *AvrLm4* and *avrLm4* isolates from the BC_5 generation, which were used to produce near-isogenic ascospores for fitness studies, were identical for more than 90% of the markers tested.

In vitro *growth rates of NI* L. maculans *isolates*

For the six avirulent (*AvrLm4*) and six virulent (*avrLm4*) NI isolates selected from the BC_5 progeny (Table 2), the *in vitro* growth rate of *avrLm4* isolates was greater than that of *AvrLm4* isolates at 4 °C ($P<0.001$; 78df; SED=0.01) and 12 °C ($P<0.01$; 80df; SED=0.02). However, these differences between *AvrLm4* and *avrLm4* NI isolates were small (0.05 mm per day at 4 °C and 0.09 mm per day at 12 °C), compared to the differences between the two parental isolates of the F_1 generation (0.18–0.36 mm per day at 4–22 °C; Table 2). In addition, there was no difference in mean growth rate between *AvrLm4* (3.62 mm per day) and *avrLm4* (3.65 mm per day) isolates at 22 °C ($P=0.13$; 60df; SED=0.03). Furthermore, there was no difference in germination patterns between ascospores of *AvrLm4* and *avrLm4* isolates on water agar at 22 °C.

Table 1. Estimation of the level of isogenicity of the two parental isolates of each generation of the back-crossing (BC) scheme for producing near isogenic (NI) *AvrLm4* and *avrLm4* isolates of *Leptosphaeria maculans*

Generation	No. AFLP markers polymorphic between parental isolates[a]	Observed % polymorphic markers[b]	Expected % polymorphic markers
F_1	94[c]	100	100
BC_1	50	53.2	50
BC_2	30	31.9	25
BC_3	15	16.0	12.5
BC_4	15	16.0	6.3
BC_5	9	9.6	3.2

[a]Parental isolates for the F1 were v23.1.3 (*MAT1–2*) and v23.1.2 (*MAT1–1*); parental isolates for BC5 were v23.1.3 (*MAT1–2*) and v37.1.4 (*MAT1–1*).
[b]Based on the proportion of AFLP markers which were polymorphic between parental isolates.
[c]A total of 94 markers were compared on each pair of parental isolates.

Development of the 'ascospore shower' method for inoculation with ascospores produced in vitro

All the plants inoculated with ascospores that had been produced on oilseed rape stem base debris under natural conditions developed leaf lesions, regardless of the inoculation method (Table 3). However, no lesions developed on unwounded plants inoculated with droplets of conidial suspensions, and very few lesions developed on plants sprayed with conidia. Plants inoculated with conidia on wounded sites developed lesions. All plants inoculated with ascospores produced *in vitro* developed lesions. The incubation period (from inoculation to appearance of the first lesion) was longer when conidia were inoculum (15–18 days) than when ascospores were inoculum (5–9 days). For ascospores, the incubation period was longer for the 'ascospore shower' method (9 days) than for droplet inoculation (5–7 days). For conidia, the incubation period was longer for spray inoculation (18 days) than for droplet inoculation (15 days).

The infection efficiency of ascospores was greater than that of conidia (Table 3). For droplet inoculation, only 56% of sites inoculated with conidia developed lesions, while 100% of sites inoculated with ascospores, either produced under natural conditions or from crosses *in vitro*, developed lesions. In treatments where leaves were sprayed with spore suspensions, about nine ascospores were required to produce one lesion, while about 2.4×10^6 conidia were necessary to produce one lesion. The 'ascospore shower' inoculation method was more efficient than the ascospore suspension method. To inoculate three plants with ascospores from *in vitro* crosses between v23.1.2 and v23.1.3, three pseudothecia were used to produce the ascospore shower and each plant developed an average of 3.5 lesions (Table 3). By contrast, ten pseudothecia were needed to make only 200 μl of ascospore suspension at the concentration required (10^3 ascospores ml^{-1}). This volume was not sufficient for spray inoculation, and could be used only for point inoculation. To inoculate three plants with ascospores produced under natural conditions, three small pieces of debris were used to produce the ascospore shower. By contrast, 10 pieces of debris were required to make an ascospore suspension for spray inoculation.

Table 2. In vitro growth rate of the *Leptosphaeria maculans* parental isolates of the F_1 and BC_5 generations and of near-isogenic isolates (*AvrLm4* and *avrLm4*) from the BC_5 progeny

Isolate[a]	AvrLm4 allele[b]	*MAT* allele[c]	Growth rate (mm day^{-1})[d]		
			4 °C	12 °C	22 °C
Parents					
v23.1.3	*AvrLm4*	*MAT1–2*	0.98	2.00	3.51
v23.1.2	*avrLm4*	*MAT1–1*	1.16	2.36	3.80
v37.1.4	*avrLm4*	*MAT1–1*	1.00	1.88	3.44
Progeny					
v41.1.2	*AvrLm4*	*MAT1–1*	1.01	2.09	3.81
v41.2.1	*AvrLm4*	*MAT1–1*	0.99	1.89	3.75
v41.5.3	*AvrLm4*	*MAT1–1*	0.99	1.78	3.59
v41.2.4	*AvrLm4*	*MAT1–2*	0.94	1.88	3.51
v41.2.5	*AvrLm4*	*MAT1–2*	1.00	1.87	3.71
v41.3.9	*AvrLm4*	*MAT1–2*	0.96	2.14	3.64
v41.1.1	*avrLm4*	*MAT1–1*	1.00	2.11	3.68
v41.2.8	*avrLm4*	*MAT1–1*	0.97	1.84	3.62
v41.3.3	*avrLm4*	*MAT1–1*	0.98	2.18	3.57
v41.3.1	*avrLm4*	*MAT1–2*	0.98	1.87	3.64
v41.3.10	*avrLm4*	*MAT1–2*	1.10	1.83	3.58
v41.5.6	*avrLm4*	*MAT1–2*	1.03	2.22	3.63
SED			0.006	0.021	0.027
(df)			(99)	(95)	(75)

[a]The parental isolates were v23.1.3 and v23.1.2 for F_1, v23.1.3 and v37.1.4 for BC_5.
[b]Isolates were characterised at the AvrLm4 locus by inoculation onto cotyledons of oilseed rape seedlings of cv Jet Neuf (with resistance gene *Rlm4*) and Westar (susceptible control with no resistance genes).
[c]The mating type (*MAT1–1* or *MAT1–2*) of isolates was determined by PCR.
[d]Radial growth of isolates grown on V8-juice agar in darkness was measured for two perpendicular diameters (mm) per plate 6, 11 and 17 days (12 and 22 °C) or 10, 24 and 32 days (4 °C) after inoculation. The growth rate was the slope of the regression line calculated from data for four times (0, 6, 11 and 17 days for 12 and 22 °C; 0, 10, 24 and 32 days for 4 °C). Data are means from three or four replicate plates per isolate.

Table 3. Effects of inoculum type (ascospores or conidia) and inoculation method on incubation period, number of lesions and infection efficiency of *Leptosphaeria maculans* (phoma leaf spot) on leaves of oilseed rape cv. Lipton

Inoculum	Inoculation method	Incubation period (days)[a]	No. lesions per plant	Infection efficiency
Conidia[b]	Point droplet	–[c]	0	0
	Wounding/point	15	4.1	56.1%[d]
	Spray droplets	18	0.6	$1/(2.4\times10^6)$[e]
Ascospores (produced naturally)[f]	Point droplet	5	5.4	100%[d]
	Spray droplets	7	45.5	1/(8.7)[e]
	Ascospore shower	9	6.5	–[c]
Ascospores (produced *in vitro*)[g]	Point droplet	7	4	100%[d]
	Ascospore shower	9	3.5	–[c]

[a]Time (days) from inoculation to appearance of the first lesion.
[b]The conidial suspension was made from a *L. maculans* isolate derived from a phoma leaf spot sampled during November 2001 from an oilseed rape (cv Apex) field experiment at Rothamsted.
[c]Not available or not tested.
[d]For point inoculation, infection efficiency is defined as the percentage of inoculated sites that developed lesions.
[e]For spray inoculation with spore suspension (conidia or ascospores), infection efficiency is defined in terms of the number of spores required to cause one lesion (e.g., 9 ascospores required to cause one lesion for spray ascospores droplets inoculation).
[f]Ascospores were obtained from naturally infested winter oilseed rape stem base debris (cv Apex) collected during August 2002 at Rothamsted.
[g]Ascospores were obtained from *in vitro* crosses between isolates v23.1.2 and v23.1.3.

Fitness of virulent (avrLm4) and avirulent (AvrLm4) L. maculans *isolates during leaf infection in controlled environments*

Oilseed rape plants inoculated with ascospores of *AvrLm4* isolates developed more lesions than plants inoculated with ascospores of *avrLm4* isolates ($P < 0.001$; 30df; SED = 0.72) (Figure 2). There was an interaction between temperature and type of inoculum ($P < 0.05$; 15df; SED = 2.2). Plants inoculated with ascospores of *AvrLm4* isolates developed the most lesions at 20 °C (10 lesions per plant) and the fewest lesions at 5 °C (only 1 lesion per plant). However, there was no significant difference in the number of lesions between temperatures on plants inoculated with ascospores of *avrLm4* isolates. There was no significant difference in the number of lesions between cultivars Darmor and Eurol; therefore only mean values are presented.

There was no difference in incubation period (from inoculation to the appearance of first lesion) between *AvrLm4* and *avrLm4* isolates at 5–25 °C ($P = 0.8$; 30df; SED = 0.2). However, there were differences between temperatures for both *AvrLm4* and *avrLm4* isolates. The relationship between incubation period (f) and temperature (T) was fitted best by a single line for *AvrLm4* and a*vrLm4* isolates: $f = 24.3–0.7\ T$ ($R^2 = 0.87$). As temperature increased, the incubation period of both *AvrLm4* and *avrLm4* isolates decreased, from 21 days at 5 °C to 8 days at 25 °C (Figure 3).

At 13–14 days after inoculation, the diameters of lesions on plants inoculated with ascospores of *AvrLm4* isolates were greater than those on plants inoculated with ascospores of *avrLm4* isolates ($P < 0.001$; 18df; SED = 0.04) (Table 4). For both *AvrLm4* and *avrLm4* isolates, the diameters of lesions increased with increased temperature ($P < 0.05$; 4df; SED = 0.1). The mean diameter of *AvrLm4* lesions increased from 3.7 mm at 15 °C to 9.8 mm at 25 °C, and that of *avrLm4* lesions increased from 3.3 mm at 15 °C to 7.1 mm at 25 °C. More lesions on plants inoculated with ascospores of *AvrLm4* isolates produced pycnidia than did lesions on plants inoculated with ascospores of *avrLm4* isolates ($P < 0.05$; 24df; SED = 1.5). On plants inoculated with ascospores of *AvrLm4* isolates, 18% of lesions produced pycnidia at 15 °C, and 11, 10 and 9% of lesions produced pycnidia at 10, 20 and 25 °C, respectively (Table 4). On plants inoculated with ascospores of *avrLm4* isolates, only 14% of lesions produced pycnidia at 15 °C, and 6, 5 and 7% of lesions produced pycnidia at 10, 20 and 25 °C, respectively. No pycnidia were observed on lesions at 5 °C. There was no difference between *AvrLm4* and *avrLm4* isolates in the percentage of lesions, that produced isolates at 1–3 cm away from lesions towards the petiole at 15–25°C (Table 4).

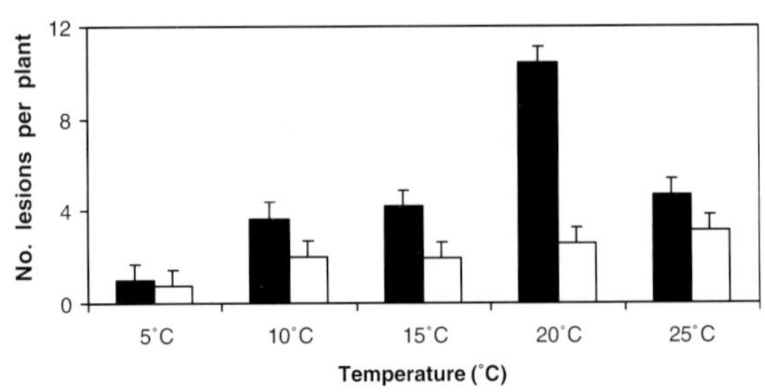

Figure 2. Numbers of phoma leaf spot lesions which developed at different temperatures on oilseed rape plants of cultivars Darmor and Eurol that lack resistance gene *Rlm4*, inoculated with ascospores of *AvrLm4* (■) or *avrLm4* (□) isolates of *Leptosphaeria maculans* produced *in vitro*. Data are combined means for cultivars from three replicate experiments ($P < 0.001$; SED = 0.72; 30df).

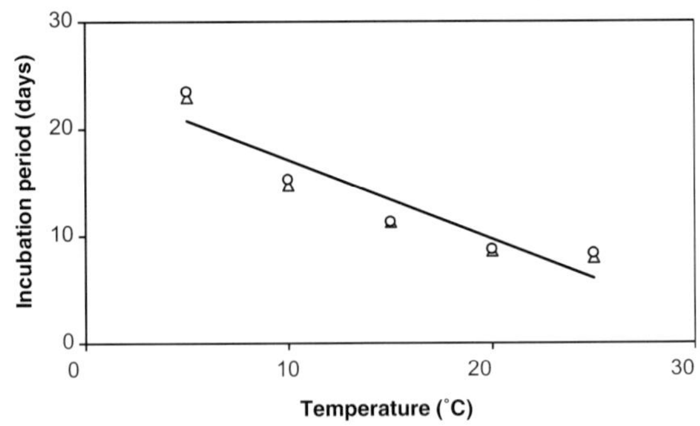

Figure 3. Effects of temperature on the incubation period (from inoculation with ascospores to the appearance of the first lesion) of *AvrLm4* (Δ) and *avrLm4* (○) isolates of *Leptosphaeria maculans* on oilseed rape plants (cultivars Darmor and Eurol) lacking the resistance gene *Rlm4*. The relationship between incubation period (f) and temperature (T) for *AvrLm4* and a*vrLm4* was fitted best by a single line $f = 24.3–0.73\ T$ ($R^2 = 0.87$). Data points are means of three replicate experiments but the regression line was calculated from data for individual experiments.

Table 4. Fitness of *Leptosphaeria maculans* near-isogenic isolates *AvrLm4* (avirulent) or *avrLm4* (virulent) in terms of lesion diameter, % lesions producing pycnidia or growth from leaf spot lesions towards the petiole on leaves of *Brassica napus* cultivars Darmor and Eurol without the corresponding resistance gene *Rlm4*[a]

Temperature (°C)	AvrLm4 allele	Lesion diameter (mm)[b]	% lesions producing pycnidia[c]	% lesions producing hyphae at 1 – 3 cm away from the lesion towards the petiole[d]		
				1 cm	2 cm	3 cm
5	*AvrLm4*	–[e]	0	–	–	–
	avrLm4	–	0	–	–	–
10	*AvrLm4*	–	10.5	–	–	–
	avrLm4	–	6.0	–	–	–
15	*AvrLm4*	3.7	17.5	78.3	61.3	58.0
	avrLm4	3.3	14.0	75.0	66.0	70.6
20	*AvrLm4*	8.8	9.5	83.3	66.2	60.1
	avrLm4	6.0	4.7	75.7	39.2	46.6
25	*AvrLm4*	9.8	9.4	64.7	53.9	39.6
	avrLm4	7.1	6.8	41.2	23.8	22.4
SED		0.04	1.56	15.38	16.61	16.60
(df)		(18)	(24)	(10)	(10)	(14)

[a]Data are the mean values for cultivars Darmor and Eurol from three replicate experiments. In each experiment, plants were inoculated with fresh ascospores of *AvrLm4* or *avrLm4* isolates produced *in vitro* on a new set of crosses.
[b]Lesion diameters were measured 13–14 days after inoculation.
[c]Number of lesions which produced pycnidia was counted at the end of each experiment.
[d]Presence of *L. maculans* hyphae shown by isolations, made on affected leaves 18–19 days after inoculation.
[e]Not measured, since lesions were too small to measure at 10 °C and no lesions had developed at 5 °C 13–14 days after inoculation.

Fitness of virulent (avrLm4) and avirulent (AvrLm4) isolates in natural populations of L. maculans

During the 2002/2003 growing season, 284 isolates were recovered from leaf lesions in autumn, 49 isolates were recovered from stem cankers in summer and 171 isolates were recovered from ascospores produced on stem debris collected from the field at Versailles. These 504 isolates were classified as *AvrLm4* or *avrLm4*. During the growing season, the frequency of *AvrLm4* isolates increased from 5.7% at the phoma leaf spot stage in autumn to 20.4% at the stem canker stage before harvest and 15.2% on debris after harvest (Figure 4). A similar trend was observed during 2003/2004, when 280 isolates (139 from leaf lesions in autumn and 141 from leaf lesions in spring) were recovered from leaf lesions, 166 from stem cankers and 244 from ascospores produced on stem debris. These 690 isolates were classified as *AvrLm4* or *avrLm4*. The frequency of *AvrLm4* isolates increased from 7.9% at the phoma leaf spot stage (in autumn/spring) to 11.5% at the stem canker stage before harvest and 12.3% on debris after harvest.

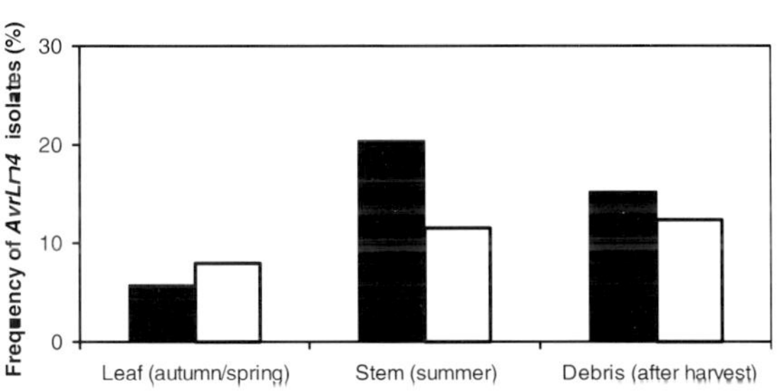

Figure 4. Frequency (%) of the avirulent allele (*AvrLm4*) in natural populations of *Leptosphaeria maculans* sampled from oilseed rape cultivars lacking the resistance gene *Rlm4* during 2002/2003 (■) and 2003/2004 (□) field experiments at Versailles, France from phoma leaf spots [sampled during autumn (2002/2003) or autumn and spring (2003/2004)], stem cankers (summer) and stem debris (after harvest).

Discussion

Results of controlled-environment experiments, indicating that *AvrLm4* and *avrLm4 L. maculans* isolates differ for some, but not all, components of phoma leaf spot development, suggest that there is a fitness cost of virulence. Firstly, plants inoculated with ascospores of *AvrLm4* isolates produced more lesions than those inoculated with ascospores of *avrLm4* isolates at all five temperatures tested, suggesting that infectivity of *AvrLm4* ascospores is greater than that of *avrLm4* ascospores. Secondly, there was no difference between *AvrLm4* and *avrLm4* isolates in the incubation period from inoculation to lesion development. Similarly, there was no difference between *AvrLm4* and *avrLm4* isolates in growth rate within the leaf within 3 cm from the infection site at 15–25 °C, as for growth on V8-agar medium. This suggests that fitness differences between avirulent and virulent isolates in field conditions may not be related to the speed of colonization of the leaf tissues. Thirdly, at all temperatures tested, *AvrLm4* ascospores produced larger leaf spot lesions than *avrLm4* ascospores and more lesions initiated by *AvrLm4* ascospores produced pycnidia than did lesions initiated by *avrLm4* ascospores. The larger leaf spot lesions may reflect a more extensive colonization of the plant tissue at the infection site by avirulent isolates, or a greater ability of these isolates to induce plant cell death and necrosis. These larger lesions may allow such isolates to benefit from a larger source of nutrients for subsequent systemic colonization of the petiole and the stem and/or sporulation.

Results from the field experiments during two seasons support the conclusions from controlled-environment experiments that there is a fitness cost of virulence. In both growing seasons, there was an increase in the frequency of *AvrLm4* isolates between the autumn (leaf spot stage) and summer (stem base canker and post-harvest stages). These data indicate that *AvrLm4* isolates may be fitter than *avrLm4* isolates in systemic growth from leaf lesions to stems under natural conditions. Growth from leaf lesions to stems is a crucial stage in the life cycle of *L. maculans* (Fitt et al., 2006). Experiments under both controlled and field conditions therefore suggest that the *avrLm4* allele may have a fitness cost associated with loss of function of the avirulence gene. Such a fitness cost of the *avrLm4* allele might explain why the cv. Jet Neuf, with the resistance gene *Rlm4*, has been successfully used as a source of resistance for 10–15 years. This also might explain why *AvrLm4* isolates are still present in field populations in France (Balesdent et al., 2006) and other parts of Europe (Stachowiak et al., 2006), although *Rlm4* has been used widely in past cultivars and is still present in more than 25% of current commercial French cultivars (Rouxel et al., 2003b; MH Balesdent, INRA and X Pinochet, CETIOM, France, unpublished data).

A pathogen "cost of virulence" parameter is often a component in models of factors affecting durability of host resistance (Pietravalle et al., 2006). This parameter is based on the hypothesis that loss of pathogen functions linked with the loss of avirulence will decrease the competitiveness of virulent isolates as compared to avirulent isolates. However, this hypothesis had not been demonstrated experimentally for a fungal plant pathogen, although one bacterium–plant interaction has been analysed (Vera-Cruz et al., 2000). Our paper therefore outlines the first experimental evidence that loss of one avirulence function in a fungal pathogen can alter its ability to produce leaf symptoms and sporulate on lesions, as compared to an avirulent isolate. Loss of the *AvrLm4* function also may affect the ability of the virulent isolates to compete with avirulent isolates during growth down the petiole to reach the stem base and to colonize the basal plant tissues in natural field conditions. This work, using both the novel ascospore shower inoculation method and NI isolates, will now be expanded to two other avirulence genes, *AvrLm1* and *AvrLm6*, to evaluate if this is a feature specific to *AvrLm4* or a more general feature of avirulence genes in *L. maculans*.

The development of near-isogenic (NI) isolates of *L. maculans* suggests that the back-crossing (BC) scheme is an efficient procedure for increasing isogenicity between *AvrLm4* and *avrLm4* isolates, since less than 10% of the AFLP markers that were polymorphic between parental isolates at the beginning of the BC scheme were still polymorphic in the fifth BC generation. In theory, the frequency of polymorphism should be halved at each BC generation, but the observed decrease in polymorphism at each generation was less than expected. This result may be linked to the need to maintain polymorphism at another genomic region (the *MAT* locus), independent of *AvrLm4*, to allow subsequent fertile crosses. The increase in isogenicity of *AvrLm4* and *avrLm4* isolates with BC also was shown by the changes in *in vitro* growth rate of the isolates. There were larger differences in *in vitro* growth rate between the parental isolates (15%) of the F_1 generation at 4 and 12 °C than between parental isolates (4%) of the fifth BC generation. These results not only illustrate the efficiency of the inbreeding process, but also indicate that these NI isolates are appropriate for fitness studies, because quantitative differences between avirulent and virulent isolates during host infection and colonization will not be a simple consequence of major differences in general growth of the isolates, but will show differences in the pathogenic fitness of the isolates. Furthermore, the crossing of NI isolates *in vitro* offers a method to produce ascospores of NI *L. maculans* for use as inoculum to compare virulent and avirulent isolates for their ability to infect and colonize plants by natural means.

As an alternative to the production of NI isolates by classical back-crossing, genetically modified isolates resulting from the transformation of a virulent isolate by the cloned avirulent allele can be used. However, it is then necessary to first clone the avirulent allele, which is often difficult, as for *AvrLm1* in *L. maculans* (Attard et al., 2002; L. Gout, INRA, France, unpubl.) and other fungi (Farman and Leong, 1998). In addition, the random integration of the cloned fragment may produce a deleterious effect on the isolate, as random integration of alien DNA is also a method for identifying mutants with decreased pathogenicity, for pathogens such as *L. maculans* (Meyer et al., 2004). Therefore, such an approach necessitates analyzing a number of transformed isolates to obtain useful results.

The 'ascospore shower' method is an efficient procedure for inoculation with *L. maculans* ascospores when inoculum is limited; for instance, ascospores produced from *in vitro* crosses. Three times as many pseudothecia were required to inoculate the same number of plants with ascospore suspension as were required with the 'ascospore shower' method. Furthermore, when using the 'ascospore shower' method, it was not necessary to wound leaves, whereas inoculation with conidial suspensions without wounding the leaves resulted in few lesions. As stem canker epidemics are initiated by air-borne ascospores, this method more accurately simulates natural conditions. Thus, it is suitable for investigating the first stages of interactions between *Brassica* species and *L. maculans*. This new 'ascospore shower' method can be combined with NI *L. maculans* isolates to investigate the fitness of other pairs of virulent and avirulent isolates, and thus the potential durability of novel resistance genes (Brun et al., 2001; Leach et al., 2001; Rouxel et al., 2003a; Sprague et al., 2006).

Acknowledgements

We thank the European Union (SECURE, QLRT–2001–01813), the UK Department for Environment, Food and Rural Affairs (OREGIN, AR0703), the Biotechnology and Biological Sciences Research Council, the Chadacre Agricultural Trust and the Royal Society for supporting this work. We thank A. D. Todd for statistical analyses of the data.

References

Attard A, Gout L, Gourgues M, Kuhn ML, Schmit J, Laroche S, Ansan-Melayah D, Billault A, Cattolico L, Balesdent MH and Rouxel T (2002) Analysis of molecular markers genetically linked to the *Leptosphaeria maculans* avirulence gene *AvrLm1* in field populations indicates a highly conserved event leading to virulence on *Rlm1* genotypes. Molecular Plant–Microbe Interactions 15: 672–682

Ansan-Melayah D, Balesdent MH, Delourme R, Pilet ML, Tanguy X, Renard M and Rouxel T (1998) Genes for race-specific resistance against blackleg disease in *Brassica napus* L. Plant Breeding 117: 373–378

Balesdent MH, Attard A, Ansan-Melayah D, Delourme R, Renard M and Rouxel T (2001) Genetic control and host range of avirulence towards *Brassica napus* cultivars Quinta and Jet Neuf in *Leptosphaeria maculans*. Phytopathology 91: 70–76

Balesdent MH, Attard A, Kühn ML and Rouxel T (2002) New avirulence genes in the phytopathologenic fungus *Leptosphaeria maculans*. Phytopathology 92: 1122–1133

Balesdent MH, Louvard K, Pinochet X and Rouxel T (2006) A large-scale survey of races of *Leptosphaeria maculans* occurring on oilseed rape in France. European Journal of Plant Pathology 114: 53–65

Biddulph JE, Fitt BDL, Leech PK, Welham SJ and Gladders P (1999) Effects of temperature and wetness duration on infection of oilseed rape leaves by ascospores of *Leptosphaeria maculans* (stem canker). European Journal of Plant Pathology 105: 769–781

Brun H, Levivier S, Ruer D, Somda I, Chèvre AM and Renard M (2000) A field method for evaluating the potential duability of new resistance sources: application to the *Leptosphaeria maculans/Brassica napus* pathosystem. Phytopathology 90: 961–966

Brun H, Ruer D, Levivier S, Somda I, Chèvre AM and Renard M (2001) Presence in *Leptosphaeria maculans* populations of isolates virulent on resistance introgressed into *Brassica napus* from the *B. nigra* B genome. Plant Pathology 50: 69–74

Cozijnsen AJ and Howlett BJ (2003) Characterisation of the mating-type locus of the plant pathogenic ascomycete *Leptosphaeria maculans*. Current Genetics 43: 351–357

Crow JF (1986) Basic Concepts in Population, Quantitative Evolutionary Genetics, Academic, New York, 273

Delourme R, Pilet-Nayel ML, Archipiano M, Horvais R, Tanguy X, Rouxel T, Brun H, Renard M and Balesdent MH (2004) A cluster of major specific resistance genes to *Leptosphaeria maculans* in *Brassica napus*. Phytopathology 94: 578–583

Delourme R, Chèvre A-M, Brun H, Rouxel T, Balesdent MH, Dias JS, Salisbury PA, Renard M and Rimmer SR (2006) Major gene and polygenic resistance to *Leptosphaeria maculans* in oilseed rape (*Brassica napus*). European Journal of Plant Pathology 114: 41–52

Farman ML and Leong SA (1998) Chromosome walking to the AVR1-CO39 avirulence gene of *Magnaporthe grisea*: discrepancy between the physical and genetic maps. Genetics 150: 1049–1058

Fitt BDL, Brun H, Barbetti MJ and Rimmer SR (2006) World-wide importance of phoma stem canker (*Leptosphaeria maculans* and *L. biglobosa*) on oilseed rape (*Brassica napus*). European Journal of Plant Pathology 114: 3–15

Gall C, Balesdent MH, Robin P and Rouxel T (1994) Tetrad analysis of acid phosphatase, soluble protein patterns and mating type in *Leptosphaeria maculans*. Phytopathology 84: 1299–1305

Gladders P and Musa TM (1980) Observations on the epidemiology of *Leptosphaeria maculans* stem canker in winter oilseed rape. Plant Pathology 29: 28–37

Huang YJ, Toscano-Underwood C, Fitt BDL, Todd AD, West JS, Koopmann B and Balesdent MH (2001) Effects of temperature on germination and hyphal growth from ascospores of A-group and B-group *Leptosphaeria maculans* (phoma stem canker of oilseed rape). Annals of Applied Biology 139: 193–207

Huang YJ, Toscano-Underwood C, Fitt BDL, Hu XJ and Hall AM (2003) Effects of temperature on ascospore germination and penetration of oilseed rape (*Brassica napus*) leaves by A-group or B-group *Leptosphaeria maculans* (phoma stem canker). Plant Pathology 52: 245–255

Huang YJ, Fitt BDL, Jedryczka M, Dakowska S, West JS, Gladders P, Steed JM and Li ZQ (2005) Patterns of ascospore release in relation to phoma stem canker epidemiology in England (*Leptosphaeria maculans*) and Poland (*L. biglobosa*). European Journal of Plant Pathology 111: 263–277

Kuhn ML, Gout L, Howlett BJ, Melayah D, Meyer M, Balesdent MH and Rouxel T (2006) Genetic linkage maps and genomic organization in *Leptosphaeria maculans*. European Journal of Plant Pathology 114: 17–31

Leach JE, Vera Cruz CM, Bai J and Leung H (2001) Pathogen fitness penalty as a predictor of durability of disease resistance genes. Annual Review of Phytopathology 39: 187–224

Li H, Sivasithamparam K and Barbetti MJ (2003) Breakdown of a *Brassica rapa* subsp. *sylvestris* single dominant blackleg resistance gene in *B. napus* rapeseed by *Leptosphaeria maculans* field isolates in Australia. Plant Disease 87: 752

Mengistu A, Rimmer RS and Williams PH (1993) Protocols for *in vitro* sporulation, ascospore release, sexual mating, and fertility in crosses of *Leptosphaeria maculans*. Plant Disease 77: 538–540

Meyer M, Blaise F, Rémy E, Zhou L, Tourneur C, Narcy JP, Rouxel T and Balesdent MH (2004) *Agrobacterium tumefaciens*-mediated transformation of *Leptosphaeria maculans*. International Organisation for Biological Control Bulletin 27: 179–192

Payne RW, Lane PW, Digby PGN, Harding SA, Leech PK, Morgan GW, Todd AD, Thompson R, Tunnicliffe Wilson G, Welham SJ and White RP (1993) Genstat 5 Release 3 Reference Manual, Clarendon Press, Oxford, UK

Pietravalle S, Lemarié S and van den Bosch F (2005) Durability of resistance and cost of virulence. European Journal of Plant Pathology 114: 107–116

Pringle A and Taylor JW (2002) The fitness of filamentous fungi. Trends in Microbiology 10: 475–481

Rouxel T, Penaud A, Pinochet X, Brun H, Gout L, Delourme R, Schmit J and Balesdent MH (2003a) A 10-year survey of populations of *Leptosphaeria maculans* in France indicates a rapid adaptation towards the *Rlm1* resistance gene of oilseed rape. European Journal of Plant Pathology 109: 871–881

Rouxel T, Willner E, Coudard L and Balesdent MH (2003b) Screening and identification of resistance to *Leptosphaeria maculans* (stem canker) in *Brassica napus* accessions. Euphytica 133: 219–231

Sprague SJ, Balesdent MH, Brun H, Hayden HL, Marcroft SJ, Pinochet X, Rouxel T and Howlett BJ (2006) Major gene resistance in *Brassica napus* (oilseed rape) is overcome by changes in virulence of populations of *Leptosphaeria maculans* in France and Australia. European Journal of Plant Pathology 114: 33–40

Stachowiak A, Olechnowicz J, Jedryczka M, Rouxel T, Balesdent MH, Happstadius I, Gladders P, Latunde-Dada A and Evans N (2006) Frequency of avirulence alleles in field populations of *Leptosphaeria maculans* in Europe. European Journal of Plant Pathology 114: 67–75

Sylvester-Bradley R and Makepeace RJ (1985) Revision of a code for stages of development in oilseed rape (*Brassica napus* L.). Aspects of Applied Biology: Field trials methods and data handling 10: 395–400

Toscano-Underwood C, West JS, Fitt BDL, Todd AD and Jedryczka M (2001) Development of phoma lesions on oilseed rape leaves inoculated with ascospores of A-group or B-group *Leptosphaeria maculans* (stem canker) at different temperatures and wetness durations. Plant Pathology 50: 28–41

Vera Cruz CM, Bai J, Oña I, Leung H, Nelson RJ, Mew T-W and Leach JE (2000) Predicting durability of a disease resistance gene based on an assessment of the fitness loss and epidemiological consequences of avirulence gene mutation. Proceedings of the National Academy of Sciences of the USA 97: 13500–13505

West JS, Biddulph JE, Fitt BDL and Gladders P (1999) Epidemiology of *Leptosphaeria maculans* in relation to forecasting stem canker severity on winter oilseed rape in the UK. Annals of Applied Biology 135: 535–546

West JS, Kharbanda PD, Barbetti MJ and Fitt BDL (2001) Epidemiology and management of *Leptosphaeria maculans* (phoma stem canker) on oilseed rape in Australia, Canada and Europe. Plant Pathology 50: 10–27

West JS, Balesdent MH, Rouxel T, Narcy JP, Huang YJ, Roux J, Steed JM, Fitt BDL and Schmit J (2002) Colonisation of winter oilseed rape tissues by A/Tox^{+} and B/Tox^{0} *Leptosphaeria maculans* (phoma stem canker) in France and England. Plant Pathology 51: 311–21

European Journal of Plant Pathology (2006) 114: 91–106
DOI 10.1007/s10658-005-3628-z

Improved resistance management for durable disease control: A case study of phoma stem canker of oilseed rape (*Brassica napus*)

J.N. Aubertot[1,*], J.S. West[2], L. Bousset-Vaslin[3], M.U. Salam[4], M.J. Barbetti[5] and A.J. Diggle[6]
[1]*UMR d'Agronomie INRA/INA P-G, BP 1, 78850, Thiverval-Grignon, France;* [2]*Rothamsted Research, AL5 2JQ, Harpenden, UK;* [3]*UMR BIO3P INRA/Agrocampus Rennes, Domaine de la Motte, BP 35327, 35653, Le Rheu Cedex, France;* [4]*Department of Agriculture, Western Australia, Centre for Cropping Systems, P.O. Box 483, 6401, Northam, WA, Australia;* [5]*School of Plant Biology, The University of Western Australia, 35 Stirling Highway, 6009, Crawley, WA, Australia;* [6]*Department of Agriculture, Western Australia, Baron-Hay Court, 6151, South Perth, WA, Australia; *Author for correspondence (Fax: +33-130815425; E-mail: aubertot@grignon.inra.fr)*

Accepted 20 September 2005

Key words: durability, Integrated Pest Management, *Leptosphaeria maculans*, population genetics, resistance breakdown, specific resistance

Abstract

Specific resistance loci in plants are generally very efficient in controlling development of pathogen populations. However, because of the strong selection pressure exerted, these resistances are often not durable. The probability of a resistance breakdown in a pathosystem depends on the evolutionary potential of the pathogen which is affected by: (i) the type of resistance (monogenic and/or polygenic), (ii) the type of reproduction of the pathogen (sexual and/or asexual), (iii) the capacity of the pathogen for dispersal, (iv) the resistance deployment strategy (pyramiding of specific resistances, mixture of cultivars, spatio-temporal alternation), (v) the size of the pathogen population, which is affected by control methods and environmental conditions. We propose the concept of Integrated Avirulence Management (IAM) to enhance the durability of specific resistances. IAM involves a strategy to limit the selection pressure exerted on pathogen populations and, at the same time, reduce the size of pathogen populations by combining cultural, physical, biological or chemical methods of control. Several breakdowns of resistance specific to *Leptosphaeria maculans*, the causal agent of phoma stem canker have occurred in Europe and in Australia. This review paper examines control methods to limit the size of *L. maculans* populations and discusses how this limitation of population size can enhance the durability of specific resistances. It proposes pathways for the development of a spatially explicit model to define IAM strategies. Simulation results are presented to demonstrate the potential uses of such a model for the oilseed rape/*L. maculans* pathosystem.

Abreviations: GM – Genetically Modified; IAM – Integrated Avirulence Management

Introduction

The durability of control methods to limit or eradicate populations of living organisms is an important, generic issue in human societies. In the domain of human health, the development of resistance to antibiotics has been well documented and studied in bacteria. In the field of crop protection, the phenomenon of adaptation of pest populations to a means of control has been widely studied for chemical control against weeds, fungi, and insects (de Waard, 1993), including development of resistance in insects against genetically modified (GM) crops (Laxminarayan and

Simpson, 2002). Lastly, genetic control of fungal, bacterial and viral plant pathogens can also lack durability (McDonald and Linde, 2002; Parlevliet, 2002).

It is therefore important to develop methods to preserve the efficacy of available genetic or chemical means to control pathogen populations. For crop protection, models have been developed to enhance the durability of pest control methods. In addition to specific breeding strategies, such as pyramiding (McDonald and Linde, 2002), several approaches have been proposed to enhance the durability of resistance to plant pathogens. These generally consist of (i) changing the proportion of resistant cultivars in a given region using non-spatially explicit models (Kiyosawa, 1982; Shi-Mai, 1991; van den Bosch and Gilligan, 2003), (ii) using mixtures of cultivars (Wolfe, 1984; Mundt, 2002), or (iii) spatio-temporal deployment of different resistances (Kiyosawa, 1982; Holt and Chancellor, 1999). Similarly, for GM crops, various models have been proposed to help define strategies to limit the selection pressure exerted on the pest populations through (i) the adaptation of refuge size (Laxminarayan and Simpson, 2002; Linacre and Thompson, 2004), (ii) refuge size and spatial organisation (Caprio, 2001; Vacher et al., 2003), (iii) refuge size, spatial organisation and crop rotation (Peck et al., 1999; Onstad et al., 2001). Computerised simulations of the evolution of insect resistance to insecticides indicated that insecticide rate, refuge size and rotation of insecticides with different modes of action could enhance the durability of the efficacy of chemical control (Gazzoni, 1998). However, these models usually take into account only one control method. Nonetheless some models aimed at defining management strategies for a non-durable method of control do integrate additional control methods. For GM crops, Storer et al. (2003) and Cerda and Wright (2004) integrated refuge size, spatial organisation, crop rotation and pesticide use on the refuge in their models to help preserve the efficacy of GM crops. In addition, non-chemical controls (such as tillage, fallow periods and sowing rate) have been generally integrated into models to develop strategies for management of herbicide resistance (Cavan et al., 2000; Neve et al., 2003). Likewise, Integrated Resistance Management (IRM) programs have been successfully developed to prevent the adaptation to insecticides of tobacco whitefly (*Bemisia tabaci*) and other pests (Palumbo et al., 2001).

With regard to the durability of genetic resistances to pathogens, fewer integrated approaches have been conducted, no doubt because of the complexity of the pathosystems. One idea proposed is the integrated management of cultivar resistance group simultaneously with fungicide application (Crute, 1984; Wolfe, 1984). For example, different powdery mildew resistance sources could be used for spring and winter barley, while permitting only specific fungicides on each crop type to isolate pathogen populations on the two barley types and hence limit evolutionary potential. Wolfe (1984) extended the idea to alter crop cultivars grown each year (in a cultivar mixture rather than with different cultivars in different fields), simultaneously altering either the chemical used or the cultivar protected by it. Where different diversification groups are available, the groups in use may be rotated each season so that any one particular group is initially unprotected for one or two seasons, then protected by a fungicide, before being omitted from cultivation completely for at least one season.

Understanding the selection occurring in a local pathogen population is complex as it is influenced by sub-populations on wild host plants, spore release from old debris (of previous cultivars), and long-distance immigration of spores or seed-borne inoculum in addition to inoculum produced on the range of cultivars currently grown in a region. Due to local differences in the pathogen population, classification of cultivar resistance for advising growers each season is therefore usually based on trials done in several locations within a particular country or region. However, classification of 'field resistance' is a combination of different components: qualitative, race-specific resistance; quantitative polygenic resistance, disease escape and tolerance, which may each be affected by cultural and environmental factors. Various cropping practices can decrease disease severity or increase 'field resistance' and therefore reduce selection of a pathogen population virulent on the current cultivars. Integrated management of pathogens can increase the durability of the resistances used for their control (McDonald and Linde, 2002; Mundt et al., 2002; Parlevliet, 2002), but there is a general lack of knowledge on how to integrate several methods to enhance the durability of resistances.

Therefore, it is important to develop tools to integrate methods to limit the selection pressure exerted on the populations by resistances and to limit the inoculum size of pathogen populations. We propose the expression "Integrated Avirulence Management" (IAM) to describe this concept.

IAM involves not only limiting the selection pressure exerted by specific resistances on pathogen populations using a gene deployment strategy, but also limiting the inoculum size of the pathogen populations using an appropriate combination of control methods. These methods can be cultural, physical, biological or chemical. The gene deployment strategy can comprise alternating genetic resistances in time and space, or pyramiding resistances into a single cultivar. The deployment strategy can be applied at plant level (pyramiding), at field level (cultivar mixtures, species mixtures, multi-lines and multi-blends), at the farm level (inter-field diversification), or at the regional level (inter-field diversification, Adugna, 2004). With regard to the deployment in time, resistance genes can be released sequentially (a new resistance is introduced as soon as resistance breakdown occurs), or different genetic resistances can be used in different seasons. The physical area to which the definition of IAM applies depends on the capacity of the pathogen for dispersal. Although this integrated management is proposed for specific resistances, it can be applied to quantitative resistances.

We illustrate the concept of IAM using phoma stem canker (*Leptosphaeria maculans*), one of the most serious diseases that affects oilseed rape worldwide (Fitt et al., 2006). The main control of this disease is genetic through specific and quantitative resistances (Delourme et al., 2006). McDonald and Linde (2002) proposed a model that predicts the risk of specific resistance breakdown for several pathosystems on a 4–11 scale, with higher values indicating greater risk of resistance breakdown. With this model, the *Leptosphaeria maculans*/oilseed rape pathosystem was scored as 7–8. This high level of risk of specific resistance breakdown results from the evolutionary potential of the pathogen that undergoes a combination of sexual and asexual reproduction and has a medium genotypic flow (West et al., 2001). The high score given by this model is consistent with observations of several specific resistance breakdowns that have occurred in Europe (Rouxel et al., 2003) and, more recently, Australia (Li et al., 2003a,b, 2004; Sprague et al., 2006). This review paper analyses the methods (cultural, physical, chemical, and biological) that can limit the size of *Leptosphaeria maculans* populations, then demonstrates how this limitation can enhance the durability of specific resistances in the considered pathosystem. Finally, it proposes pathways for the development of a spatially explicit model to define IAM strategies and provides preliminary simulation results.

Cultural, physical, chemical and biological control of Leptosphaeria maculans

Cropping practices affecting the severity of phoma stem canker include crop rotation, stubble management and tillage, sowing date, fungicide regime, plant density and fertiliser use. The first group (crop rotation, stubble management and tillage) concerns ways of reducing exposure of crops to inoculum. Similarly, sowing date can be changed to avoid exposure to air-borne spores completely or at least during the period when the crop is most vulnerable. Sowing date may also be changed to avoid excessively hot periods, which can stress plants and exacerbate the impact of the disease (West et al., 2001). Fungicide regime is included here as an agricultural practice that when used correctly can substantially reduce final disease severity, but which for some regions is not an economic option. Other factors such as plant density and soil nutrition affect the severity of disease in various ways through disease escape or increasing crop tolerance.

Cultural and physical control

Reducing exposure of the crop to inoculum by disposal of stubble and using crop rotation to maximise separation between recent debris and the current crop is an important strategy for disease control. Similarly, tillage regime has a large effect on inoculum availability through differences in the burial or destruction of debris bearing pathogen fruiting bodies. This not only influences dispersal gradients of air-borne ascospores but also prevents conidia being rain-splashed directly from stubble in close proximity to plants (Thürwächter et al., 1999). The most common primary inoculum is air-borne ascospores, which are usually released after rain over several weeks, from pseudothecia formed on remains of infected plants.

Tillage regimes tested in Canada by Turkington et al. (2000) confirm that burial helps the breakdown of debris but that the woody root and crown debris is more persistent than upper stem debris. After 3 years, residue levels were generally low following all tillage regimes tested (including zero-tillage), indicating that a rotation with at least 4 year intervals is advisable to reduce phoma stem canker in that location. This contrasts to the situation in south-eastern Australia where crop residues and ascospore loads decrease substantially after 18 months (Marcroft et al., 2003).

Typically, ascospore dispersal gradients follow a negative exponential (or a power function, or a Cauchy) distribution with most spores deposited within 500 m of the source, especially the first 100 m, but with some risk to crops up to several kilometres away (West et al., 2001; Marcroft et al., 2004). However, dispersal gradients are modified substantially by wind speed and local topography. Although the small number of spores likely to travel considerable distances pose little risk of initiating severe disease in the current season, these spores could represent new races or pathotypes, which can then become established in a region. The dispersal and deposition of many air-borne particles, such as plant pathogen spores and plant spores or pollens, has been modelled to optimise separation distances. This approach can be applied to calculate the risk of nearby inoculum sources for infection of oilseed rape crops by ascospores of *L. maculans*. Further work is required to quantify the relative proportions of background air-borne spores that have travelled long distances and those produced locally.

Currently, a separation of crops from stubble sources of only 500 m is recommended in Australia. This distance is short because current cultivars are generally more resistant than those grown in the 1970s, use of minimum tillage practices is widespread and the acreage of oilseed rape grown is increasing, making large separation distances from stubble sources difficult (Barbetti and Khangura, 1999). Due to the dispersal characteristics of ascospores, and the separation distances recommended, the size of crop areas may also act as a cultural method to reduce disease in combination with crop rotation. In the UK, there has been a tendency in recent years for several 'traditional farms' to be managed as one, often with the same crops being grown in adjacent fields, although cultivars with different maturities are used to lengthen the duration of the harvest and to diversify disease resistance genes. Further work is required to assess the impact of rotating relatively large crop areas compared to smaller areas. Although larger cultivated areas increase the size of sources of ascospores (from the previous season's crop debris), separation distances between the current and previous season's crops are also larger, causing dilution of spores in the atmosphere. This is generally likely to decrease disease incidence, but increase spatial heterogeneity in disease severity.

The timing of ascospore release makes sowing date crucial. Poisson and Pérès (1999) demonstrated that early leaves (i.e. the first two true leaves) are more readily infected than later leaves (leaves 4 or 6), although phoma leaf spots may take longer to appear on early leaves. Previous work has shown that infection of cotyledons or leaves early in the season produces severe phoma stem cankers before harvest (Hammond and Lewis, 1986; Poisson and Pérès, 1999; Aubertot et al., 2004a). In some areas, such as the dry Mediterranean climate of Western Australia, where the first rains of the season often synchronise the release of air-borne ascospores with seedling emergence, severe canker or even death results as the stem of seedlings up to the six leaf stage may be completely severed at the base (Barbetti and Khangura, 1999). Similar symptoms have been reported on spring rape in Canada (Kharbanda, 1993). In much of Western Europe, there is usually sufficient soil moisture to sow seeds in late summer to establish plants that have produced several leaves by the time *L. maculans* ascospores are first released, avoiding infection at the crop's most vulnerable stage (LePage and Penaud, 1995). This may explain why seedling death due to hypocotyl infection is rare in Europe (Paul and Rawlinson, 1992). However, if late harvesting of crops or poor weather delays sowing or emergence, crops may still be relatively young when ascospores initiate infections. Zhou et al. (1999) and Sun et al. (2001) showed that in the UK, early appearance of crown canker symptoms was associated with greatest yield loss, with a thermal time from onset of leaf spotting to canker appearance that differed between cultivars. However, an exception due to disease escape often occurs in some central and eastern European countries such as Poland, where autumn leaf infections occur but severe crown cankers are very rare. This is thought to be because infections caused by early (autumn) spore releases [in Poland] often do not reach the stem because early frosts cause leaves to drop off prematurely. The cold winter weather largely prevents further spore release, leading to a second peak of spore release in the spring, which leads to infection of upper stems.

Other agricultural practices affecting disease severity include plant (sowing) density and fertiliser regime. A dense stand tends to produce plants with longer petioles and causes shaded leaves, low in the canopy, to be shed sooner than in stands of low plant density, in which plants receive better illumination. Hence higher plant densities promote disease escape or delay the pathogen reaching the stem. Conversely dense stands and excessive fertiliser treatment can produce relatively tall, thin-stemmed plants, which ultimately may be more prone to lodging. Further research is required to understand effects of plant density on phoma stem canker severity and yield. Spring nitrogen fertilisation has no effect on phoma stem canker severity (Söchting and Verreet, 2004), but high autumn nitrogen availability (i.e. more than 250 kg ha^{-1}) increases phoma stem canker severity (Aubertot et al., 2004a).

Chemical and biological control

Chemical control of phoma stem canker includes seed treatments, coated fertiliser granules or fungicide sprays to foliage or stubble, depending on the disease epidemiology and crop economics (West et al., 2001). Foliar fungicide applications are used only where potential yields are high, such as Western Europe, rather than in areas where shorter day-lengths and heat-induced early senescence cause lower yields. Generally, foliar fungicide applications have an impact on final canker severity only if applied at the optimal time (West et al., 2002). Penaud et al. (1999) reported a system developed in France to improve fungicide use based on infection risk and agronomic factors. Imminent infection (7 rain-days after sowing, maturation of pseudothecia or first detection of >20 ascospores per day) triggers the decision to apply a fungicide if agronomic factors (cultivar susceptibility, soil type, growth stage and plant vigour) favour disease. The optimal timing for foliar fungicide applications in England is soon after the onset of significant leaf spotting (i.e. >10–20% plants affected) as this prevents new leaf infections, whilst preventing stem infection from existing leaf lesions due to fungistatic and fungicidal effects (West et al., 2002). However, this approach relies on regular disease assessment in crops, particularly if plants are small, as the incidence of leaf spotting can increase rapidly and the fungus can reach the stem of small plants relatively quickly. To reduce the need for regular assessment of crops, the onset of phoma leaf spot and therefore, fungicide timing, can now be predicted. The model developed by Salam et al. (2003) predicts the onset of pseudothecial maturity in Australia when 43 suitable days (as defined by temperature and rainfall parameters) have occurred since harvest. Following the onset of maturity, ascospore showers are predicted when daily rainfall exceeds a threshold. This model is very accurate under Western Australian conditions and is used as a decision support tool for advising sowing date. Further models suited to other regions are under development to predict ascospore release or the first appearance of phoma leaf spot from temperature and wetness or rainfall data.

Chemical treatment of stubble to inhibit sporulation or even kill the pathogen has been described previously (Humpherson-Jones and Burchill, 1982, Rawlinson et al., 1984; Petrie, 1995) and a new group of chemicals has now been tested (Wherrett et al., 2003). When dry crop debris were temporarily immersed in fluquinconazole, flutriafol (technical grade) or glyphosate-ammonium, pseudothecial maturation and subsequent ascospore discharge were delayed. A practical solution such as spraying stubble and weeds after harvest could therefore help the crop to evade infection at the most vulnerable growth stages.

Several microbiological control agents that might control phoma stem canker development have been tested. The bacterium *Erwinia herbicola* was highly antagonistic to *L. maculans in vitro* through antifungal substance secretion (Chakraborty et al., 1994). Another bacterium, *Paenibacillus polymyxa* strain PKB1, produces antifungal peptides (Beatty and Jensen, 2002). This bacterium significantly reduced germination and germ-tube length of *L. maculans* conidia *in vitro*, decreased incidence and severity of leaf lesions in growth chamber tests, and decreased survival of the fungus on infected oilseed rape stubble in field experiments (Kharbanda et al., 1999). *Cyathus striatus*, a bird's nest fungus, also has antagonistic effects against *L. maculans* via two mechanisms. It possesses a greater ability for primary resource capture (greater production of cellulase and lignase than *L. maculans*) and produces an antibiotic complex named cyathin (Maksymiak and Hall, 2000). Furthermore, a biopreparation based on spores from the fungus *Trichoderma harzianum* reduced the severity of *L. maculans* leaf lesions in field experiments (Hysek et al., 2002). However, none of these biological control agents are currently used by farmers to control *L. maculans* and the potential for their wide-scale deployment is unknown.

Size of pathogen populations and durability of resistances

Five evolutionary forces, described below, influence pathogen population evolution (McDonald and Linde, 2002) but differ in their relative importance in the oilseed rape/ *L. maculans* pathosystem in relation to durability of resistances.

Evolutionary forces that act on pathogen populations

Mutation is the ultimate source of genetic variation, leading to the appearance of new alleles in the population. In a gene-for-gene relationship between pathogen and plant host, mutation from avirulence to virulence is required in the pathogen to generate individuals able to overcome the corresponding resistance gene in the plant. Mutation rates are generally low, and vary between loci and pathogens (McDonald and Linde, 2002). However, for a given mutation rate, the larger the pathogen population, the higher the probability that virulent mutants will appear. Furthermore, as increasing numbers of virulent mutants are generated, the diversity of genetic backgrounds in which further mutations to virulence occur increases, thus increasing the probability that well adapted mutants will arise.

Genetic drift may lead to a loss of alleles over time in small populations, or populations going through regular reductions in size. However, this loss of alleles will only occur in the absence of selection. Alleles experiencing selection pressure will persist, whatever the size of the population. Drift may lead to a random loss of inessential virulence alleles on a susceptible plant cultivar but on a resistant cultivar these alleles will not be lost by drift because they are required for pathogen infection.

Migration allows the exchange of alleles between different populations. From a source of propagules, the dispersal gradient is defined as the expected number of propagules reaching a given distance. The number of propagules usually decreases with increasing distance, and, depending on the nature of propagules and the dispersive agent (wind, rain, vectors, human activities), the pathogen dispersal gradient may vary in steepness. However, for a given gradient, increasing the size of the source population will increase both the number of propagules dispersed to a given distance, and the probability that some propagules will travel further.

Recombination in plant pathogens is generally considered to promote adaptation of populations by promoting the association of several necessary alleles. If a combination of several virulence alleles is necessary to infect a cultivar, a strictly asexual pathogen will need to accumulate these virulences by successive mutations in the same genotype, whereas sexual reproduction will enable the combination of two or more alleles that appeared independently in different genotypes. In large populations, increased numbers of mutants and reduced genetic drift will promote higher diversity, and therefore increase probability for combinations of virulences to occur. However, to undergo sexual reproduction, opposite mating types of plant pathogens must meet and cross on the same plant or its residues. Therefore, recombination can occur only between those alleles that are not necessary to infect the cultivar. In other cases, either the two genotypes do not meet because they cannot infect the same plant, or they meet but both already possess the same alleles so recombination does not confer any advantage. A disadvantage of sexual reproduction is the need to find the opposite mating type in order to produce progeny. As stated above, mating types of obligatory heterothallic pathogens must meet on the same plant to cross, and are restricted to passive dispersal (they cannot look for each other as insects do). In large populations, finding the opposite mating type will not be a problem because incidence of the disease is high and most plants are infected by several genotypes. However, to overcome resistance genes, when populations are very small on the resistant cultivar, finding the opposite mating type may become a limiting factor. This problem may be reduced in *L. maculans* as saprophytic colonisation of debris by pathotypes that would be avirulent on the living plant could enhance the prospect of crosses with virulent pathotypes.

Selection for virulence, selection against virulence (fitness cost) and selection for aggressiveness are critical evolutionary forces involved in the process of resistance breakdown. In the gene-for-gene relationship, major resistance genes oblige the pathogen to possess the corresponding virulence allele. Therefore, as soon as a resistant cultivar is grown in a region, virulent genotypes have an advantage over avirulent ones, in the proportion of the respective acreages. Strategies that moderate the selection pressure by diversification of genes in space or time have been reviewed previously (Finckh et al., 2000; McDonald and Linde, 2002; Mundt, 2002). In general, the larger the acreage of a resistant cultivar, the greater the advantage of virulent genotypes - regardless of the size of pathogen populations. The effect of increasing the total pathogen population in the region (i.e. on susceptible cultivars) will be to generate a larger number of virulent genotypes through mutation, more combinations of alleles through recombination and more migration of these compatible genotypes onto the resistant cultivars, all of which will increase the size of populations on the resistant cultivars.

Selection against virulence, or a fitness cost for the virulent individuals over avirulent ones, has often been introduced into models of pathogen population evolution (Kiyosawa, 1982; Leonard, 1997; Damgaard, 1999). Without a fitness penalty, the only stable state that a model could reach is ascendancy of multi-virulent individuals, which does not correspond to what is observed in the field. The strength of this selection against virulence is still a matter of debate because of difficulty in measuring it. If one assumes that all single base pair mutations are similarly likely, then the equilibrium frequency of an allele is a simple function of its historical relative fitness. As mutations are rare, small differences in fitness should cause large differences in frequency, so often fitness costs for initially rare virulence alleles would be immeasurably small. Mutations that confer virulence against one type of resistance could lead to avirulence against another type of resistance, but may be equally fit in the absence of either. This process seems to vary among pathogens and loci and probably only applies when avirulent and virulent genotypes are in competition (i.e. on the susceptible cultivar). Furthermore, in the case of insecticide resistance in mosquitoes, where the initial fitness cost is very high, fitness is an adaptive trait and replacement by alleles with lower fitness cost occurs over time (Chevillon et al., 1999). In large pathogen populations, if the resource is limiting, competition between individuals is expected to be more severe, but on the other hand the diversity of genotypes in which mutation to virulence occurs is higher, thus increasing the probability of lower fitness costs or compensatory mutations.

Selection pressure imposed by quantitative resistance is less documented than is the case for major gene resistance (gene-for-gene relationships). In the case of quantitative resistance, infection by the pathogen is not an all or nothing response, therefore the selection pressure imposed on the pathogen population is lower and as a consequence more difficult to study. Large populations of a pathogen increase the probability of selecting genotypes with increased virulence.

Durability of resistance, evolutionary forces and population size in the oilseed rape/L. maculans *pathosystem*

What is a durable use of resistance genes, and which forces act for and against adaptation of pathogen populations? In some cases, durability of resistance occurs at the level of the gene, and the pathogen is not able to become virulent. One famous case is the barley *mlo* gene for resistance against *Blumeria graminis* for which laboratory selection of virulent mutants was unsuccessful and no virulent isolates could be recovered from crops despite about 30 years of use in spring barley (Schwarzbach, 1998). In this case, three independent mutations are necessary for the fungus to become virulent and the only two virulent isolates available suffer a fitness cost (Lyngkjaer et al., 2000). Of course, breeders hope to find such genes but for *L. maculans* such combinations of virulence genes have not been found. Furthermore, virulent *L. maculans* isolates were present even before the introduction of the corresponding genes in the cultivars. This has been the case not only for genes originating in *B. napus*, but also in France for genes transferred into *B. napus* from the mustards *B. juncea* (Brun et al., 2000) and *B. nigra* (Brun et al., 2001) or in Australia for a gene introduced from *B. rapa* subsp. *sylvestris* (Li et al., 2003b, 2005). In these cases, reducing the size of the population at the local level may reduce the occurrence of new mutants and virulence alleles may be lost through genetic drift. However, as virulent isolates are detected in populations in the complete absence of selection pressure, it appears unlikely that their fitness cost would be sufficient to hamper their build-up as soon as resistant plants were present.

In some cases individual plants may have durable resistance through combination or pyramiding of stable resistance genes. The underlying assumption of pyramiding is that it is more difficult for a pathogen to acquire several independent mutations than a single one. However, any separate use of the individual resistance genes elsewhere will reduce the efficacy of pyramiding (Kiyosawa, 1982). In the case of *L. maculans*, if virulent mutants are already present at low frequencies in the population, they would need to acquire only one allele to become virulent against two resistance genes. Isolates with multiple virulences may also be already present in the population, at a frequency below the detection threshold but would be exposed to a very strong selection pressure. Furthermore, once multi-virulent isolates are present in the population, corresponding resistance genes cannot be used either alone or in combination.

A final approach to enhance the durability of resistances occurs on a regional scale through strategic use of non-durable genes in a manner that prevents adaptation of the pathogen population, despite its ability to overcome individual genes or gene combinations. In this case what happens in each field (population of the fungus) should be distinguished from what happens at the level of a network of fields or a small region. At a regional scale, the group of fields from which the inoculum is produced at the end of one growing season is mixed to form a metapopulation of the fungus and spread over all of the crops emerging in the next growing season. On this scale, maintaining low, stable frequencies of virulent alleles will enhance durability of resistance genes. The susceptible cultivars (on which virulence alleles are neutral) act as a source of virulent mutants, and resistant cultivars (on which virulence alleles will multiply) act as a sink. As long as the resistant cultivars remain only a sink, the composition of the metapopulation will not change over time. When the resistant cultivars contribute progeny to the next generation, no combination of the processes of mutation, drift, migration, recombination or fitness cost will be sufficient to counterbalance the selective advantage of virulent individuals growing on the whole acreage in contrast to avirulent individuals that do not.

Durability in a metapopulation and cultural practices

Until now, strategies aimed at increasing durability of resistance have generally sought to reduce selection pressure by diversification of resistance genes in space and/or time. However, the expected outcome from such strategies is a dilution that slows the process, but does not result in a stable equilibrium, unless a huge fitness cost is assumed. Much less attention has been paid to other possibilities, in the context of the metapopulation, that focus on preventing resistant cultivars from becoming a source of inoculum. To be a source from a resistant cultivar, *L. maculans* has to survive saprophytically and produce inoculum, which has to be dispersed onto oilseed rape fields in following seasons. All three processes could be hampered in the following ways:

1. Survival of the pathogen is altered by various stubble management practices described previously and fungicide sprays.
2. Inoculum is produced on oilseed rape residues as either ascospores arising from sexual recombination, or conidia from asexual multiplication. As a heterothallic obligatory pathogen, *L. maculans* can produce ascospores only if mycelia of opposite mating types meet on the same plant. Adapting the crop management of oilseed rape (sowing date and rate,

nitrogen availability, fungicide application) can reduce the susceptibility and exposure of the crop to inoculum and therefore reduce the subsequent inoculum production.

3. Dispersal of inoculum from the resistant cultivar to the following year's crops depends on the spore type, the amount of inoculum and the locations of fields in successive years. Ascospores move between distant fields, whereas conidia at most will only reach adjacent fields. Therefore, unless the oilseed rape is grown in the same field each year, or sown adjacent to unburied residues from the previous season, the transmission of offspring from one generation to the next through conidia is likely to be very low. Spatio-temporal distribution of cultivars in a given region will therefore strongly affect the incidence and severity of leaf spot infections.

Thus cultural practices will play a major role in durability of resistances, if used on both resistant and susceptible cultivars so as to prevent resistant cultivars from acting as an inoculum source. On resistant cultivars, the amount of inoculum has to be reduced, and the success of dispersal constrained. The amount of inoculum produced may be lowered actively by soil tillage (Turkington et al., 2000), burning (Marcroft et al., 2003) or use of pesticides (Humpherson-Jones and Burchill, 1982; Rawlinson et al., 1984; Petrie, 1995; Wherrett et al., 2003, 2004). The success of dispersal can be constrained by making sure conidia do not survive from one season to the next (no consecutive rotation, no new crop adjacent to field residues of resistant cultivars). Concerning ascospores, the key point is to reduce the size of the population on resistant cultivars, which will both lower the distance at which migration will be significant and reduce inoculum produced if the size of the population is low enough so that individuals do not find an opposite mating type for sexual reproduction. Decreasing population size on the resistant cultivars also depends on the reduction of population size on susceptible cultivars. Smaller population sizes on susceptible cultivars are expected to reduce the number of virulent mutants, to lower the probability that such a mutation occurs in a background with low fitness costs, to enhance the loss of virulence alleles through genetic drift, and to lower allelic diversity so that fewer allelic combinations are generated by drift. Thus, smaller population sizes on susceptible cultivars reduce the probability that virulent mutants reach fields sown with otherwise resistant cultivars where they would not be prevented from obtaining an automatic selective advantage.

Modelling for Leptosphaeria maculans Integrated Avirulence Management

To apply IAM strategies, modelling is an essential tool to handle the complexity of a pluri-annual system, driven by environmental variables along with technical decisions taken by a group of stakeholders at a regional scale. In this section, pathways are presented for the development of a model aimed at defining IAM strategies. This model describes the effects of cultural practices of several farmers in a given region on the evolution of the genetic structure of the pathogen populations. To do so, cultural practices that greatly affect the pathosystem will be taken into account; changes over time in the regional spatial distribution of oilseed cultivars (with possibly different specific and quantitative resistances), stubble management (tillage, export, or burning), application of chemicals to leaves or stubble, sowing date and sowing rate, and nitrogen fertiliser inputs throughout the cropping systems (Figure 1). The distribution of infected stubble is predicted from the changes over time in regional spatial distribution of oilseed rape cultivars and stubble management. Application of chemicals to stubble may reduce and delay the production of primary inoculum. Temperature and rainfall affect the timing of pseudothecial maturity; the mature ascospores are released during a rainfall event and wind speed and direction indicate where the ascospores will be dispersed. The deposition of ascospores on oilseed rape crops also depends on crop density (affected by nitrogen availability at sowing), as long as the soil surface is not completely covered by the crop. The severity of stem canker depends on cultivar resistance, the use of foliar fungicides, sowing date and sowing rate, nitrogen availability at sowing and weather variables. Available models, such as SimCanker (Aubertot et al., 2004b), can be used to represent this particular stage in the epidemiology of phoma stem canker. The genetic structure of

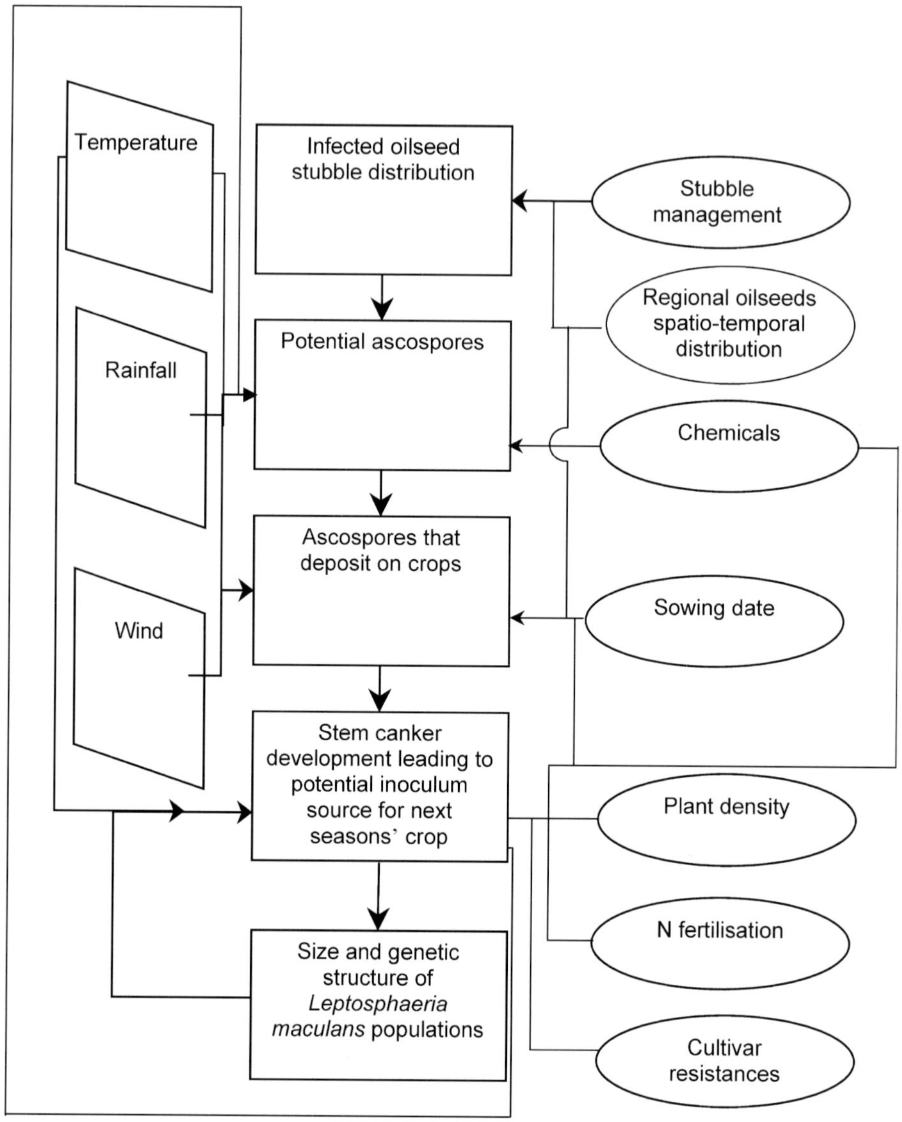

Figure 1. Flow chart of a spatially explicit model aimed to define Integrated Avirulence Management strategies for phoma stem canker of oilseed rape over several seasons.

L. maculans populations is then simulated for each field in the region studied.

Preliminary simulations illustrate how such a tool could help to define integrated strategies to enhance the durability of specific resistance genes. Simulations are limited to the spread of ascospores for one season (only the 3 upper boxes are used in these simulations). The numerical model used to compute these simulations is a combination of three validated models: a model representing the effects of tillage operations on the vertical distribution of oilseed rape stubble (Schneider, 2005); Sporacle (Salam et al., 2003); and Anthracnose Tracer (Diggle et al., 2002).

We modelled a part of Scaddan district of Western Australia (33.44°S and 121.72°E), a 17.2×17.5 km area, to indicate the pattern and intensity of ascospore dispersal. Figure 2 shows the simulated dispersal of ascospores for a period of 30 days after emergence in the current season from field 67, where an affected crop was grown the previous season. Local weather data for 1999 were used to run this simulation on an hourly basis, assuming a crop emergence date of 10th

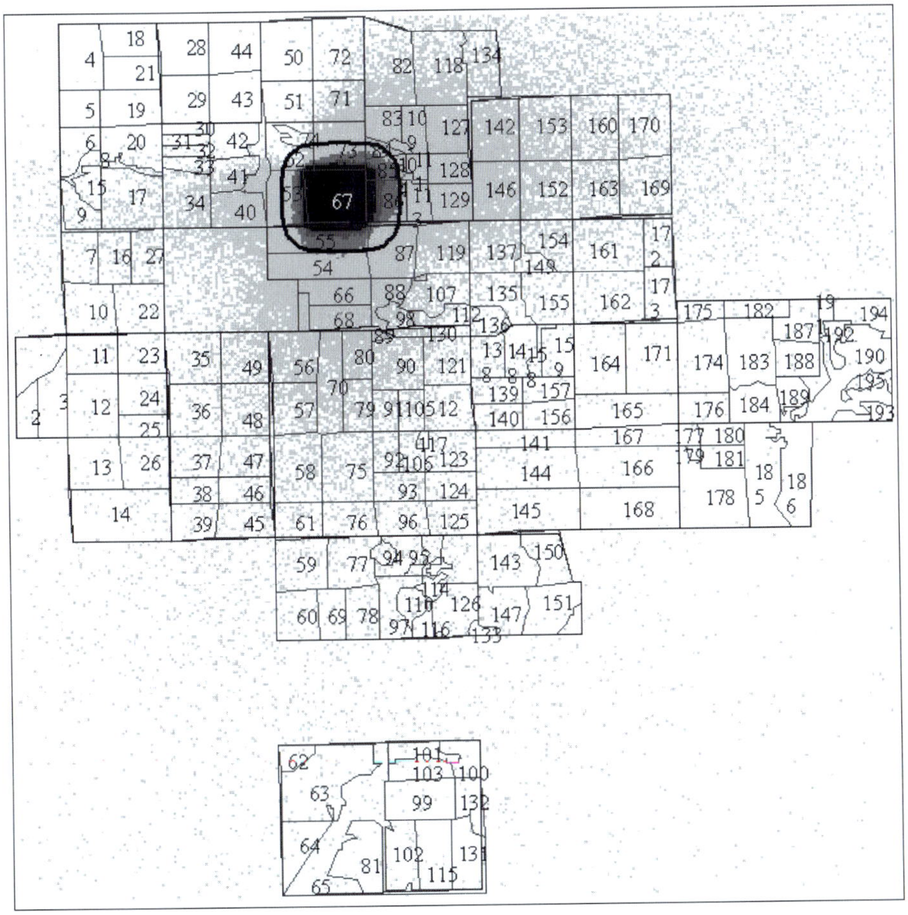

Figure 2. Relative cumulated number of *Leptosphaeria maculans* ascospores/m^2 dispersed during 30 days after emergence on a 17.2×17.5 km area of Scaddan district, Western Australia, from a source with 1.7×10^9 ascospores/m^2 (50×50 m grid). The shading represents the proportion of ascospores dispersed from sources during 30 days after emergence (□ 0%, ■ 0–1%, ■ 1–3%, ■ 3–8%, ■ >8%). The numbers correspond to each field represented on the map. The black line around field 67 illustrates the significance of the 500 m separation distance from the previous season's crop, suggested in Western Australia.

May. The functions (relationship between weather variables and ascospore maturity and seasonal pattern of ascospore release) and related parameters used have been described (Salam et al., 2003). The ascospore dispersal function has two components, the angle and distance of dispersal, as characterised by the wind attributes. The wind-induced dispersal angle is chosen from a normal distribution defined by the average wind direction, as measured by a weather station at 10 s intervals. The distance component is a random value chosen from a half-Cauchy distribution with a median dispersal parameter multiplied by the average wind speed (m s^{-1}). This parameter is the median of the distribution distance (m) that spores may travel. Details of the dispersal function have been given by Diggle et al. (2002). The dispersal parameter of Salam et al. (2001) was used to describe dispersal of *L. maculans* ascospores in Western Australia.

Wherrett et al. (2004) have recently quantified a relationship between ascospore-production and phoma stem canker severity in oilseed rape. With respect to a potential of 34×10^6 ascospores per stem (or 1.7×10^9 spores m^{-2}, considering 50 affected stem fragments m^{-2}), a disease index gradient [a calculation based on numbers of plants affected in each of 6 canker severity classes (Wherrett et al., 2004)] of 0–40, 40–50, 50–60 and >60% was observed with roughly 1, 3, 6, 8 and >8% spore-load (relative proportion of total spore production over 30 days after emergence), respectively. A similar relative scale was used to produce as outputs the concentration of ascospores dispersed in a grid (with squares 50×50 m) in five concentrations – white (no spores), light grey (0–1% of spore-load), grey (1–3% of spore-load), dark grey (3–8% of spore-load) and black (>8% of spore-load). Figure 2 also shows (black line around field 67) the significance of suggested separation distance (500 m) from the previous season's crop.

With the above assumptions, four scenarios were simulated, combining two cultural factors, cropping pattern (clustered fields and dispersed fields) and tillage (simplified or mixed) to show differences in spore dispersal in the region. For these scenarios, it was assumed that six oilseed rape crops were infected in the previous season in the region. The infected fields were either clustered (fields 51, 52, 53, 67, 71 and 73, Figure 2) or scattered (21, 25, 60, 67, 123 and 176, Figure 2). Simplified tillage consisted of sowing directly after a cover-crop operation. Mixed tillage combined three fields with conventional tillage and three fields with simplified tillage. The conventional tillage consisted of ploughing after harvest and sowing after a rotary harrow operation. The spore-load in the field with conventional tillage was 6% of that observed with the simplified tillage (from Schneider, 2005).

Figure 3 indicates that the concentration of ascospores is greatest, over the whole region, when

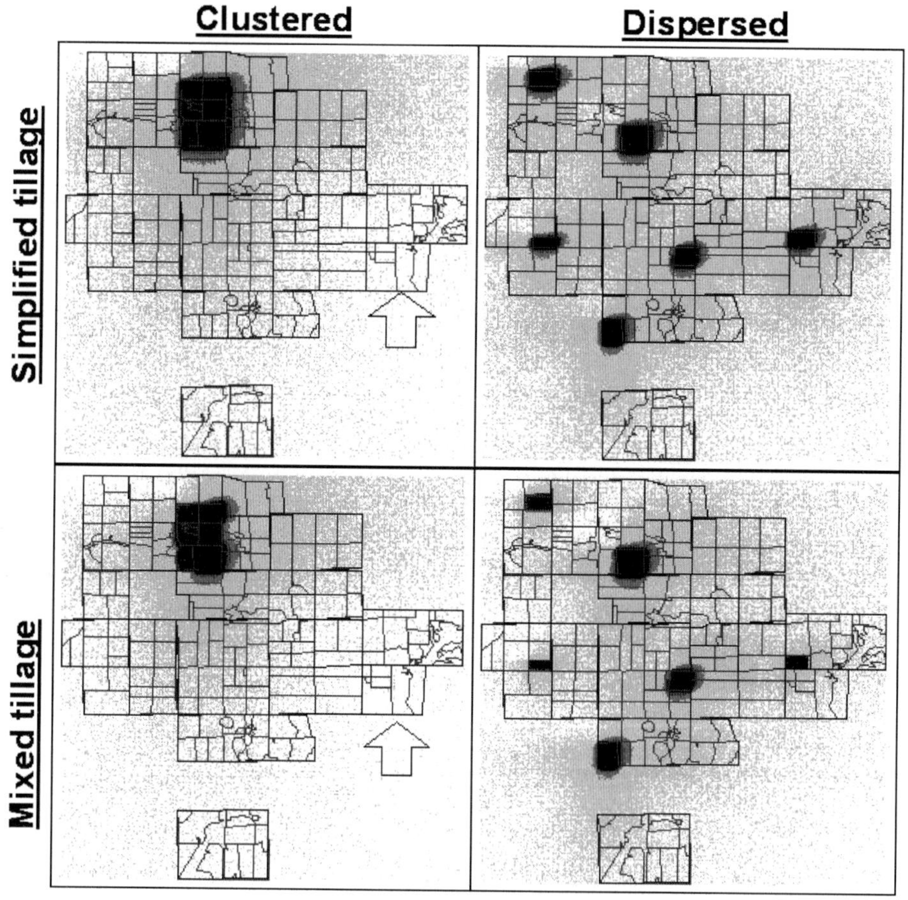

Figure 3. Comparison of the spatial dispersal of *Leptosphaeria maculans* ascospores obtained for four scenarios corresponding to six infected fields arranged in either a clustered or dispersed cropping pattern, with simplified or mixed tillage (three fields with simplified tillage and three fields with conventional tillage). The shading represents the proportion of ascospores dispersed from sources during 30 days after emergence, with a total production of 1.7×10^7 ascospores/m^2 (□ 0%, ■ 0–1%, ■ 1–3%, ■ 3–8%, ■ >8%). The arrows point to locations in the region where crops could be grown without much risk expected of ascospore deposition.

the fields are in a clustered arrangement and have undergone simplified tillage. The simplified tillage practice with a scattered crop distribution, decreases the regional ascospore concentration by 13% (compared to the greatest), whereas mixed tillage, irrespective of cropping pattern, decreases the ascospore concentration by 45% (compared to the greatest). However, the spores are expected to be more widely distributed under the dispersed (spores landed in 59 and 44% of the area, respectively for simplified or mixed tillage) compared to clustered (spores landed in 37 and 28% of the area, respectively for simplified or mixed tillage) cropping patterns. It is evident from Figure 3 that in some locations in the region, crops could be grown without much risk expected of ascospore deposition (near arrow) under the clustered cropping pattern; this is not the case for the dispersed cropping pattern scenario.

Discussion

Oilseed rape is facing a dynamic situation where there are continual changes in the *L. maculans* population, with the host being exposed to potentially threatening 'metapopulations'. While strategies for enhancing durability of resistance have existed for some time, the potential for exploiting a more integrated management approach to increase durability has largely gone untapped, mainly due to inadequate knowledge on how to integrate available control strategies for enhancing durability of resistance. Strategies and options will vary in different regions of the world. A sound approach would be to apply an integrated approach to limit the selection pressure on the *L. maculans* population and limit the actual amount of inoculum produced by the *L. maculans* population (IAM concept). However, this type of management can be successfully achieved only by including the full array of available cultural practices to decrease residue/inoculum levels (e.g. by tillage, burning, crop rotation, application of chemicals to residues), avoid residue/inoculum (e.g. time of sowing, crop rotation), and judicious and timely application and/or delivery to the crop of fungicides and fertilisers. This requires spatially explicit modelling, especially as additional factors that influence *L. maculans* population evolution, such as mutation, genetic drift, migration, recombination and selection occurring within the pathogen populations, also need to be taken into consideration in relation to durability of the host resistance gene(s). Therefore, further work is required to transform the conceptual model presented in this paper into a completely validated simulation model to provide recommendations for strategies to be followed by farmers. The integrated disease management strategy required differs in relation to the resistance genes present in the crop. For example, the integrated management of cultivars with polygenic versus major gene resistance is likely to differ between different farms and regions. Previous models have generally focused only on one or at most a few of these aspects affecting durability of host resistance. In contrast, our model is the first that attempts to spatially accommodate the full range of the key aspects affecting durability of resistance, especially those associated with cultural control, with the aim of predicting the best options for maintaining and extending the durability of host resistance against *L. maculans*. Durability of host resistance can be achieved only by combining all these approaches rather than merely relying solely upon the development and deployment of resistant cultivars.

Acknowledgements

We are grateful for funding from the French Institut National de la Recherche Agronomique through the grant "Impact des innovations variétales", the Australian Grains Research and Development Corporation, The University of Western Australia, and the UK Biotechnology and Biological Sciences Research Council and Department for the Environment, Food and Rural Affairs.

References

Adugna A (2004) Alternate approaches in deploying genes for disease resistance in crop plants. Asian Journal of Plant Sciences 3: 618–623

Aubertot JN, Pinochet X and Doré T (2004a) The effects of sowing date and nitrogen availability during vegetative stages on *Leptosphaeria maculans* development on winter oilseed rape. Crop Protection 23: 635–645

Aubertot JN, Pinochet X, Reau R and Doré T (2004b) SimCanker: a simulation model for containing phoma stem canker of oilseed rape through cultural practices. Proceedings of the 4th International Crop Science Congress. 26 September–1 October. Brisbane, Australia. http://www.regional.org.au/au/cs/2004/

Barbetti MJ and Khangura RK (1999) Managing blackleg in the disease-prone environment of Western Australia. Proceedings of the 10th International Rapeseed Congress, September 26–29, Canberra, Australia. http://www.regional.org.au/au/gcirc/

Beatty PH and Jensen SE (2002) *Paenibacillus polymyxa* produces fusaricidin-type antifungal antibiotics active against *Leptosphaeria maculans*, the causative agent of blackleg disease of canola. Canadian Journal of Microbiology 48: 159–169

Brun H, Levivier S, Somda I, Ruer D, Renard M and Chèvre AM (2000) A field method for evaluating the potential durability of new resistance sources: Application to the *Leptosphaeria maculans – Brassica napus* pathosystem. Phytopathology 90: 961–966

Brun H, Ruer D, Levivier S, Somda I, Chèvre AM and Renard M (2001) Presence in *Leptosphaeria maculans* populations of isolates virulent on resistance introgressed into *Brassica napus* from *B. nigra* B genome. Plant Pathology 50: 69–74

Caprio MA (2001) Source-sink between transgenic and non-transgenic habitats and their role in the evolution of resistance. Journal of Economic Entomology 94: 698–705

Cavan G, Cussans J and Moss SR (2000) Modelling different cultivation and herbicide strategies for their effect on herbicide resistance in *Alopecurus myosuroides*. Weed Research 40: 561–568

Cerda H and Wright DJ (2004) Modeling the spatial and temporal location of refugia to manage resistance in *Bt* transgenic crops. Agriculture, Ecosystems and Environment 102: 163–174

Chakraborty BN, Chakraborty U and Basu K (1994) Antagonism of *Erwinia herbicola* towards *Leptosphaeria maculans* causing blackleg disease of *Brassica napus*. Letters in Applied Microbiology 18: 74–76

Chevillon C, Raymond M, Guillemaud T, Lenormand T and Pasteur N (1999) Population genetics of insecticide resistance in the mosquito *Culex pipiens*. Biological Journal of the Linnean Society 68: 147–157

Crute IR (1984) The integrated use of genetic and chemical methods for control of lettuce downy mildew (*Bremia-Lactucae* Regel). Crop Protection 3: 223–241

Damgaard C (1999) Coevolution of a plant host-pathogen gene-for-gene system in a metapopulation model without cost of resistance or cost of virulence. Journal of Theoretical Biology 201: 1–12

Delourme R, Chèvre AM, Brun H, Rouxel T, Balesdent M, Dias JS, Salisbury P, Renard M and Rimmer SR (2006) Major gene and polygenic resistance to *L. maculans* in oilseed rape (*Brassica napus*). European Journal of Plant Pathology 114: 41–52

de Waard MA (1993). Management of resistance to pesticides. In: Zadoks JC (ed.) Modern Crop Protection: Development and Perspectives (pp 53–60) Wageningen Pers, Wageningen, The Netherlands

Diggle AJ, Salam MU, Thomas GJ, Yang HA, O'Connell M and Sweetingham MW (2002) Anthracnose Tracer: A spatiotemporal model for simulating the spread of anthracnose in a lupin field. Phytopathology 92: 1110–1121

Finckh MR, Gacek ES, Goyeau H, Lannou C, Merz C, Mundt CC, Munk L, Nadziak J, Newton AC, deVallavieille-Pope C and Wolfe MS (2000) Cereal variety and species mixtures in practice, with emphasis on disease resistance. Agronomie 20: 813–837

Fitt BDL, Brun H, Barbetti MJ and Rimmer SR (2006) World-wide importance of phoma stem canker (*Leptosphaeria maculans* and *L. biglobosa*) on oilseed rape (*Brassica napus*). European Journal of Plant Pathology 114: 3–15

Gazzoni DL (1998) Modeling insect resistance to insecticides using Velvetbean Caterpillar (*Anticarsia gemmatalis*) as an example. Pesticide Science 53: 109–122

Hammond KE and Lewis BG (1986) The timing and sequence of events leading to stem canker disease in populations of *Brassica napus* var. *oleifera* in the field. Plant Pathology 35: 551–564

Holt J and Chancelor TCB (1999) Modelling the spatio-temporal deployment of resistant varieties to reduce the incidence of rice tungro disease in a dynamic cropping system. Plant Pathology 48: 453–461

Humpherson-Jones FM and Burchill RT (1982) Chemical suppression of the sexual stage of *Leptosphaeria maculans* on oil-seed rape and turnip seed crop straw. Annals of Applied Biology 100: 281–288

Hysek J, Vach M, Brozova J, Sychrova E, Civinova M, Nedelnik J and Hruby J (2002) The influence of the application of mineral fertilizers with the biopreparation supresivit (*Trichoderma harzianum*) on the health and the yield of different crops. Archives of Phytopathology and Plant Protection 35(2): 115–124

Kharbanda PD (1993) Blackleg of Canola in Alberta: Investigations on Biology, Epidemiology and Management. Alberta Environmental Centre, Vegreville, AB. AECV93-R5. 86 pp

Kharbanda PD, Yang J, Beatty P, Jensen S and Tewari JP (1999) Biocontrol of *Leptosphaeria maculans* and other pathogens of canola with *Paenibacillus polymyxa* PKB1. Proceedings of the 10th International Rapeseed Congress. Canberra, Australia. http://www.regional.org.au/au/gcirc/

Kiyosawa S (1982) Genetics and epidemiological modeling of breakdown of plant disease resistance. Annual Review of Phytopathology 20: 93–117

Laxminarayan R and Simpson RD (2002) Refuge strategies for managing pest resistance in transgenic agriculture. Environmental and Resource Economics 22: 521–536

Leonard KJ (1997). Modelling gene frequency dynamics. In: Crute IR, Holub EB and Burdon JJ (eds.) The Gene-for-Gene Relationship in Plant-Parasite Interactions (pp 211–230) CAB International, London

LePage R and Penaud A (1995) Tout se joue avec le premier pic d'ascospores. CETIOM – Oléoscope 28: 23–27

Li H, Barbetti MJ and Sivasithamparam K (2003a) Responses of *Brassica napus* cultivars to *Leptosphaeria maculans* field isolates from Western Australia. Brassica 5: 25–34

Li H, Sivasithamparam K and Barbetti MJ (2003b) Breakdown of a *Brassica rapa* ssp. *sylvestris* single dominant blackleg resistance gene in *B. napus* rapeseed by *Leptosphaeria maculans* field isolates in Australia. Plant Disease 87: 752

Li H, Damour L, Sivasithamparam K and Barbetti MJ (2004) Increased virulence and physiological specialization among Western Australian isolates of *Leptosphaeria maculans* breaking down existing single dominant gene-based resistance in six cultivars of *Brassica napus*. Brassica 6: 9–16

Li H, Barbetti MJ and Sivasithamparam K (2005) Hazard from reliance on cruciferous hosts as sources of major gene based resistance for managing blackleg (*Leptosphaeria maculans*) disease. Field Crops Research 91: 185–191

Linacre NA and Thompson CJ (2004) Dynamics of insect resistance in *Bt*-corn. Ecological Modelling 171: 271–278

Lyngkjaer MF, Newton AC, Atzema JL and Baker SJ (2000) The barley *mlo*-gene: An important powdery mildew resistance source. Agronomie 20: 745–756

Maksymiak MS and Hall AM (2000) Biological control of *Leptosphaeria maculans* (anamorph *Phoma lingam*) causal agent of blackleg/canker on oil seed rape by *Cyanthus striatus*, a bird's nest fungus. The BCPC Conference: Pests and Diseases, Volume 1. Proceedings of an International Conference held at the Brighton Hilton Metropole Hotel, Brighton, UK, 13–16 November, pp. 507–510

Marcroft SJ, Sprague SJ, Pymer SJ, Salisbury PA and Howlett BJ (2003) Factors affecting production of inoculum of the blackleg fungus *Leptosphaeria maculans* in south-eastern Australia. Australian Journal of Experimental Agriculture 43: 1231–1236

Marcroft SJ, Sprague SJ, Pymer SJ, Salisbury PA and Howlett BJ (2004) Crop isolation, not extended rotation length, reduces blackleg (*Leptosphaeria maculans*) severity of canola (*Brassica napus*) in south-eastern Australia. Australian Journal of Experimental Agriculture 44: 601–606

McDonald BA and Linde C (2002) The population genetics of plant pathogens and breeding strategies for durable resistance. Euphytica 124: 163–180

Mundt CC (2002) Use of multiline cultivars and cultivar mixtures for disease management. Annual Review of Phytopathology 40: 381–410

Mundt CC, Cowger C and Garrett K (2002) Relevance of integrated management to disease durability. Euphytica 124: 245–252

Neve P, Diggle AJ, Smith FP and Powles SB (2003) Simulating evolution of glyphosate resistance in *Lolium rigidum* I: Population biology of a rare resistance trait. Weed Research 43: 404–417

Onstad DW, Guse CA, Spencer JL, Levine E and Gray ME (2001) Modeling the dynamics of adaptation to transgenic corn by western corn rootworm (*Coleoptera: Chrysomelidae*). Journal of Economic Entomology 94: 529–540

Palumbo JC, Horowitz AR and Prabhaker N (2001) Insecticidal control and resistance management for *Bemisia tabaci*. Crop Protection 20: 739–765

Parlevliet JE (2002) Durability of resistance against fungal, bacterial and viral pathogens: present situation. Euphytica 124: 147–156

Paul VH and Rawlinson CJ (1992) Diseases and Pests of Rape, Verlag Th. Mann Gelsenkirchen-Buer, Germany

Peck SL, Gould F and Ellner SP (1999) Spread of resistance in spatially extended regions of transgenic cotton: Implications for management of *Heliothis virescens* (Lepidoptera: Noctuidae). Journal of Economic Entomology 92: 1–16

Penaud A, Bernard C, Maisonneuve C, Pérès A and Pilorgé E (1999) Decision rules for a chemical control of *Leptosphaeria maculans*. Proceedings of the 10th International Rapeseed Congress, September 26–29, Canberra, Australia. http://www.regional.org.au/au/gcirc/

Petrie GA (1995) Effects of chemicals on ascospore production by *Leptosphaeria maculans* on blackleg-infected canola stubble in Saskatchewan. Canadian Plant Disease Survey 75: 45–50

Poisson B and Pérès A (1999) Study of rapeseed susceptibility to primary contamination of *Leptosphaeria maculans* in relation to plant vegetative stage. Proceedings of the 10th International Rapeseed Congress, September 26–29, Canberra, Australia. http://www.regional.org.au/au/gcirc/

Rawlinson CJ, Muthyalu G and Cayley GR (1984) Fungicide effects on light leaf-spot, canker, crop growth and yield of winter oil-seed rape. Journal of Agricultural Science 103: 613–628

Rouxel T, Penaud A, Pinochet X, Brun H, Gout L, Delourme R, Schmit J and Balesdent MH (2003) A ten-year survey of populations of *Leptosphaeria maculans* in France indicates a rapid adaptation towards the *Rlm1* resistance gene of oilseed rape. European Journal of Plant Pathology 109: 871–881

Salam MU, Galloway J, Khangura RK, Diggle AJ, MacLeod WJ and Barbetti MJ (2001) Spatial spread of Blackleg in canola – A regional scale simulation model. Conference Proceedings, 12th Australian Research Assembly on Brassicas, Geelong, Victoria, Australia, pp. 101–105

Salam MU, Khangura RK, Diggle AJ and Barbetti MJ (2003) Blackleg Sporacle: A model for predicting onset of pseudothecia maturity and seasonal ascospore showers in relation to blackleg of canola. Phytopathology 93: 1073–1081

Schneider O (2005) Analyse des effets du mode de gestion des résidus de colza sur l'initiation du cycle de *Leptosphaeria maculans* (Desm.) Ces et de Not. PhD thesis Institut National Agronomique Paris-Grignon, 93 pp

Schwarzbach E (1998) The *mlo* based resistance of barley to mildew and the response of mildew populations to the use of varieties with the *mlo* gene. Czech Journal of Genetics and Plant Breeding 34: 3–10

Shi-Mai Z (1991) PANCRIN, a prototype model of the pandemic cultivar-race interaction of yellow rust on wheat in China. Plant Pathology 40: 287–295

Söchting HP and Verreet JA (2004) Effects of different cultivation systems (soil management, nitrogen fertilization) on the epidemics of fungal diseases in oilseed rape (*Brassica napus* L. var. *napus*). Zeitschrift Fur Pflanzenkrankheiten Und Pflanzenschutz-Journal of Plant Diseases and Protection 111: 1–29

Sprague SJ, Balesdent MH, Brun H, Hayden HL, Marcroft SJ, Pinochet X, Rouxel T and Howlett BJ (2006) Major gene resistance in *Brassica napus* (oilseed rape) is overcome by changes in virulence of populations of *Leptosphaeria maculans* in France and Australia. European Journal of Plant Pathology 114: 33–40

Storer NP, Peck SL, Gould F, Van Duyn JW and Kennedy GG (2003) Sensitivity analysis of a spatially-explicit stochastic simulation model of the evolution of resistance in *Helicoverpa zea* (Lepidoptera: Noctuidae) to *Bt* trangenic corn and cotton. Journal of Economic Entomology 96: 173–187

Sun P, Fitt BDL, Steed JM, Underwood CT and West JS (2001) Factors affecting development of phoma canker (*Leptosphaeria maculans*) on stems of winter oilseed rape (*Brassica napus*) in southern England. Annals of Applied Biology 139: 227–242

Thürwächter F, Garbe V and Hoppe HH (1999) Ascospore discharge, leaf infestation and variations in pathogenicity as criteria to predict impact of *Leptosphaeria maculans* on oilseed rape. Journal of Phytopathology 147: 215–222

Turkington TK, Clayton GW, Klein-Gebbinck H and Woods DL (2000) Residue decomposition and blackleg of canola: Influence of tillage practices. Canadian Journal of Plant Pathology 22: 150–154

Vacher C, Bourguet D, Rousset F, Chevillon C and Hochberg ME (2003) Modelling the spatial configuration of refuges for a sustainable control of pests: A case study of *Bt* cotton. Journal of Evolutionary Biology 16: 378–387

van den Bosch F and Gilligan CA (2003) Measures of durability of resistance. Phytopathology 93: 616–625

West JS, Kharbanda PD, Barbetti MJ and Fitt BDL (2001) Epidemiology and management of *Leptosphaeria maculans* (phoma stem canker) on oilseed rape in Australia, Canada and Europe. Plant Pathology 50: 10–27

West JS, Fitt BDL, Leech PK, Biddulph JE, Huang Y-J and Balesdent MH (2002) Effects of timing of *Leptosphaeria maculans* ascospore release and fungicide regime on phoma leaf spot and phoma stem canker development on winter oilseed rape (*Brassica napus*) in southern England. Plant Pathology 51: 454–463

Wherrett AD, Sivasithamparam K and Barbetti MJ (2003) Chemical manipulation of *Leptosphaeria maculans* (blackleg disease) pseudothecial development and timing of ascospore discharge from canola (*Brassica napus*) residues. Australian Journal of Agricultural Research 54: 837–848

Wherrett AD, Sivasithamparam K and Barbetti MJ (2004) Establishing the relationship of ascospore loads with blackleg (*Leptosphaeria maculans*) severity on canola (*Brassica napus*). Australian Journal of Agricultural Research 55: 849–854

Wolfe MS (1984) Trying to understand and control powdery mildew. Plant Pathology 33: 451–466

Zhou Y, Fitt BDL, Welham SJ, Gladders P, Sansford CE and West JS (1999) Effects of severity and timing of stem canker (*Leptosphaeria maculans*) symptoms on yield of winter oilseed rape (*Brassica napus*) in the UK. European Journal of Plant Pathology 105: 715–728

European Journal of Plant Pathology (2006) 114: 107–116
DOI 10.1007/s10658-005-3479-7

Durability of resistance and cost of virulence

Stéphane Pietravalle[1,*], Stéphane Lemarié[2] and Frank van den Bosch[1]
[1]*Rothamsted Research, Harpenden, Hertfordshire, AL5 2JQ, UK; Current address: Central Science Laboratory, Sand Hutton, York Y041 1LZ, UK;* [2]*Unité Mixte de Recherche GAEL, INRA – Université Pierre Mendès France, BP 47, 38040, Grenoble Cedex 9, France;* **Author for correspondence (Phone: +44-1904-462672; Fax: +44-1904-462111; E-mail: s.pietravalle@csl.gov.uk)*

Accepted 7 September 2005

Key words: model complexity, ordinary differential equations, polycyclic epidemic, seasonality

Abstract

A seasonal model, where a growing season is defined as the time between sowing and harvest and alternates with an inter-crop period, was derived to study the effects of the 'cost of virulence' and cropping ratio on durability of resistance. We assumed a single strain of virulent pathogen, a single strain of avirulent pathogen and two cultivars (one resistant and one susceptible) and studied two measures of durability of resistance ('take-over time' and 'usefulness time'). Take-over time is defined as the time needed for the virulent strain of the pathogen to reach a preset threshold and predominate over the previous pathogen population. Usefulness time is the time needed before the estimated gain in green canopy area duration per plant through the use of the resistant cultivar becomes negligible. The model suggested that, although it could take several seasons before the virulent strain of the pathogen predominated over the previous pathogen population, the usefulness time of the resistant cultivar was always much shorter. Furthermore, increasing selection for the virulent strain of the pathogen (through increasing the cropping ratio of the resistant cultivar) caused the virulent strain of the pathogen to invade the system more rapidly. Cost of virulence, reflecting differences in pathogen infection rates between the four possible combinations of cultivar/pathogen strain, significantly affected durability of resistance, with the dynamics of the virulent and avirulent strains ranging from a case where the virulent strain of the pathogen died out to a case where the virulent strain of the pathogen invaded the resident pathogen population. An intermediate state, where the system reached equilibrium and the virulent strain of the pathogen neither became predominant nor died out, was defined as 'coexistence' of both strains of the pathogen. Occurrence of coexistence was directly related to the cost of virulence since it did not occur when virulence of the pathogen did not have a fitness cost. Two methods to include cost of virulence in the model gave similar results in relation to the two measures of durability of resistance studied.

Introduction

Growth in the productivity of agriculture during the last 50 years has relied, to a large extent, on the development of strategies for crop protection (pesticides, resistant cultivars or specific farming practices). It is generally recognized that both pesticides and cultivar resistance face major durability problems (Regev, 2002). Once such a crop protection technique is introduced, it is often only a matter of time before it loses its efficiency as a growing proportion of the targeted pest/pathogen becomes resistant (e.g., to fungicides). This problem can be managed as long as farmers have access to a wide range of alternative efficient crop protection techniques. However, the recent introduction of regulations for market approval has greatly decreased the number of commercialized active ingredients used in pesticides. In future, farmers may need to rely on a narrow range of chemical crop protection techniques and consequently will need to depend more on genetic resistance. Therefore, understanding and developing strategies for managing the durability of resistance in cultivars is a scientific issue with major economic stakes.

This problem of durability is particularly important in the case of the pathogen *Leptosphaeria maculans* that causes phoma stem canker in oilseed rape. Currently, the control of the disease through the use of fungicides is frequently only partially effective (Aubertot et al., 2006) and farmers often rely on cultivar resistance and specific cultural practices. Under some circumstances, major gene resistance used in cultivars has broken down shortly after its introduction into commercial cultivars. For example, the major resistance genes *Rlm1* and *Rlm6* were rendered ineffective within, respectively, 4–5 years and 3 years in France (Rouxel et al., 2003) and the major resistance gene in cultivar Surpass 400 was broken down in 3 years in Australia (Li et al., 2003).

Although mapping work has been done on *L. maculans* avirulence genes (Kuhn et al., 2006), the genetic basis of much of the resistance is not well understood in oilseed rape (Delourme et al., 2006) and strategies to maximize durability of resistance in commercial crops remain poorly developed and have mainly concentrated on crop rotation (Marcroft et al., 2002), burning or disposal of stubble (Marcroft et al., 2003) or ploughing (Turkington et al., 2000). Known major genes for resistance to *L. maculans* are scarce. In Europe, of the 10 such genes (*Rlm*) known in oilseed rape (Howlett, 2004), many have already been overcome. As a result, correct advice on optimum strategies for deployment of resistant cultivars is crucial to maximize durability of resistance.

However, because of the cost and time needed, few field experiments to study resistance durability have been done (Brun et al., 2000) and, in some situations, they may not be feasible. For instance, pyramiding and non-pyramiding of major genes of resistance have never been compared in field experiments, although pyramiding has been recently suggested as a way to improve durability of oilseed rape resistance to *L. maculans* (Li et al., 2004a). Models can be used to provide plant pathologists with a better insight into the relative effects of different agricultural practices and characteristics of the pathosystem on durability of resistance. Agricultural practices and strategies which can be investigated include crop rotation and spatial deployment of resistance genes (McDonald and Linde, 2002). However, before examining such strategies in detail, it is necessary to define a *base-line model*, discuss the assumptions it relies on and establish its validity as a starting point for further analysis. This is the aim of this paper. Hence, the emphasis of this work is on qualitative (rather than quantitative) analysis.

Van den Bosch and Gilligan (2003) developed a continuous model for a polycyclic disease epidemic and compared several ways to define durability of resistance. However, as they acknowledge in that paper, "the model used [...] is the simplest possible model that shows the effects of population dynamics and population genetics on durability" and does not include spatial nor temporal patterns. To incorporate spatial and temporal heterogeneity, it is necessary to introduce seasonality into the model. Here, we discuss differences caused by the introduction into the model of seasonality, with a sowing and a harvest date, when studying a polycyclic disease. When comparing models for monocyclic epidemics and polycyclic epidemics, we had found no significant difference in the effects studied (Pietravalle et al., 2004), possibly because a polycyclic epidemic can be considered as a series of successive monocyclic epidemics. Consequently, we decided to restrict this study to polycyclic epidemics as this led to a simpler model. Furthermore, and most importantly, we also introduced a cost of virulence. The aims of this paper are therefore (1) to explain the relationship between cropping ratio and durability of resistance, (2) to explain the relationship between cost of virulence and durability of resistance, (3) to compare two measures of durability of resistance and (4) to compare two models for introducing cost of virulence.

Materials and methods

The model

We consider a pathosystem with two host cultivars (resistant or susceptible) and two strains of the pathogen (virulent or avirulent). As well as introducing seasonality into the model, we also include two definitions of cost of virulence to represent the differences in 'total' infection rate between the four possible combinations of pathogen strain and cultivar. Further, in the context of this model, we include both qualitative resistance (complete resistance, mediated by a single gene) and quantitative resistance (incomplete resistance, mediated by one or more genes). Thus, where resistance is incomplete, it is possible for the avirulent pathogen to infect the resistant cultivar. These assumptions lead to a system of six linked ordinary differential equations (ODE) to describe the system within each growing season and a set of rules for the transfer of the pathogen and crop densities between seasons. Two equations describe the dynamics of the uninfected host, according to its susceptibility (resistant or susceptible), whereas four equations describe the dynamics of the pathogen, according to its virulence (virulent or avirulent) and the cultivar it infects (resistant or susceptible). We now derive the two models studied and define the two measures of durability used in this paper.

Within season dynamics

The densities of healthy tissue on susceptible and resistant cultivars are called S and R, respectively. In the absence of disease, both these densities increase logistically as the crop grows, between sowing and harvest. However, the increase in both healthy crop densities is slowed down or even reversed because of pathogen infection. Similarly, when present, each strain of the pathogen increases as a result of new infections and decreases as a result of death or cessation of sporulation. This is expressed as a set of six differential equations:

$$\frac{dS}{dt} = [\text{crop growth}]_S - \begin{bmatrix}\text{disease due to the}\\ \text{virulent strain}\end{bmatrix} - \begin{bmatrix}\text{disease due to the}\\ \text{avirulent strain}\end{bmatrix}$$

$$\frac{dR}{dt} = [\text{crop growth}]_R - \begin{bmatrix}\text{disease due to the}\\ \text{virulent strain}\end{bmatrix} - \begin{bmatrix}\text{disease due to the}\\ \text{avirulent strain}\end{bmatrix}$$

$$\frac{dA_S}{dt} = \begin{bmatrix}\text{disease due to avirulent}\\ \text{spores transferred from}\\ \text{susceptible to susceptible cultivar}\end{bmatrix} + \begin{bmatrix}\text{disease due to avirulent}\\ \text{spores transferred from}\\ \text{resistant to susceptible cultivar}\end{bmatrix} - \begin{bmatrix}\text{pathogen death}\\ \text{or sporulation}\\ \text{cessation}\end{bmatrix}$$

$$\frac{dA_R}{dt} = \begin{bmatrix}\text{disease due to avirulent}\\ \text{spores transferred from}\\ \text{susceptible to resistant cultivar}\end{bmatrix} + \begin{bmatrix}\text{disease due to avirulent}\\ \text{spores transferred from}\\ \text{resistant to resistant cultivar}\end{bmatrix} \begin{bmatrix}\text{pathogen death}\\ \text{or sporulation}\\ \text{cessation}\end{bmatrix}$$

$$\frac{dV_S}{dt} = \begin{bmatrix}\text{disease due to virulent}\\ \text{spores transferred from}\\ \text{susceptible to susceptible cultivar}\end{bmatrix} + \begin{bmatrix}\text{disease due to virulent}\\ \text{spores transferred from}\\ \text{resistant to susceptible cultivar}\end{bmatrix} - \begin{bmatrix}\text{pathogen death}\\ \text{or sporulation}\\ \text{cessation}\end{bmatrix}$$

$$\frac{dV_R}{dt} = \begin{bmatrix}\text{disease due to virulent}\\ \text{spores transferred from}\\ \text{susceptible to resistant cultivar}\end{bmatrix} + \begin{bmatrix}\text{disease due to virulent}\\ \text{spores transferred from}\\ \text{resistant to resistant cultivar}\end{bmatrix} - \begin{bmatrix}\text{pathogen death}\\ \text{or sporulation}\\ \text{cessation}\end{bmatrix}$$

In the absence of disease, healthy crop densities of both resistant and susceptible cultivars increase logistically, with equal growth rate r ($r = r_R = r_S$) and with carrying capacities K_R and K_S. The pathogen has a polycyclic development with a base-line infection rate β (defined as the new area infected per lesion per unit time for the avirulent strain of the pathogen on the susceptible crop) and a rate of pathogen death and sporulation cessation μ. We also introduce a cost of virulence that affects the pathogen infection rate. To do so, we define parameters ε_1, ε_2 and ε_3 that represent the relative decreases in infection rates of the virulent pathogen on the susceptible crop, the avirulent pathogen on the resistant crop (possibly to zero) and virulent pathogen on the resistant crop, respectively. The parameters ε vary between 0 (no infection possible) and 1 (no decrease in infection rate). We also need to introduce the number (g) of lesions produced per infected area. In this study, to investigate the four objectives presented in the 'Results' section, we chose to keep the eight parameters ($r, K, \beta, \mu, g, \varepsilon_1, \varepsilon_2$ and ε_3) constant across growing seasons.

In a first approach, we assume that the cost of virulence ε affects infection and does not affect spore production (model 1) (Pietravalle et al., 2004). Within growing season, this leads, after setting $A = A_S + A_R$ and $V = V_S + V_R$, to a set of four differential equations, with parameters described in Table 1.

$$\begin{aligned}
\frac{dS}{dt} &= r_S S\left(1 - \frac{S}{K_S}\right) - \beta SA - \varepsilon_1 \beta SV \\
\frac{dR}{dt} &= r_R R\left(1 - \frac{R}{K_R}\right) - \varepsilon_2 \beta RA - \varepsilon_3 \beta RV \\
\frac{dA}{dt} &= g\beta A(S + \varepsilon_2 R) - \mu A \\
\frac{dV}{dt} &= g\beta V(\varepsilon_1 S + \varepsilon_3 R) - \mu V
\end{aligned} \qquad (1)$$

In a second approach, we assume the cost of virulence ε affects spore production and does not affect infection (model 2). This leads to six differential equations:

$$\begin{aligned}
\frac{dS}{dt} &= r_S S\left(1 - \frac{S}{K_S}\right) - \beta S(\varepsilon_1 V_S + \varepsilon_3 V_R) \\
&- \beta S(A_S + \varepsilon_2 A_R) \\
\frac{dR}{dt} &= r_R R\left(1 - \frac{R}{K_R}\right) - \beta R(\varepsilon_1 V_S + \varepsilon_3 V_R) \\
&- \beta R(A_S + \varepsilon_2 A_R) \\
\frac{dA_S}{dt} &= g\beta(A_S + \varepsilon_2 A_R)S - \mu A_S \\
\frac{dA_R}{dt} &= g\beta(A_S + \varepsilon_2 A_R)R - \mu A_R
\end{aligned}$$

Table 1. Variables and parameters used in the models

	Description	Dimension
Variables		
R	Density of uninfected resistant crop	L^2L^{-2}
S	Density of uninfected susceptible crop	L^2L^{-2}
A_R	Density of avirulent lesions on resistant crop	L^{-2}
A_S	Density of avirulent lesions on susceptible crop	L^{-2}
V_R	Density of virulent lesions on resistant crop	L^{-2}
V_S	Density of virulent lesions on susceptible crop	L^{-2}
Parameters		
r_R, r_S	Logistic growth rate of resistant or susceptible crop	t^{-1}
K_R, K_S	Carrying capacity of resistant or susceptible crop	L^2L^{-2}
g	Number of lesions produced per infected area	L^{-2}
μ	Rate of pathogen death and sporulation cessation	t^{-1}
β[a]	New area infected per lesion per unit time (base-line)	L^2t^{-1}
$\varepsilon_{1,2,3}$[a]	Cost of virulence	1
λ	Rate of transfer of pathogen between seasons	1
θ	Proportion of crop sown which is resistant	1

[a]New area infected per lesion per unit time for the avirulent pathogen on the susceptible and resistant crops are β and $\varepsilon_2\,\beta$ respectively and new area infected per lesion per unit time for the virulent pathogen on the susceptible and resistant crops are $\varepsilon_1\,\beta$ and $\varepsilon_3\,\beta$ respectively.

$$\frac{dV_S}{dt} = g\beta(\varepsilon_1 V_S + \varepsilon_3 V_R)S - \mu V_S$$
$$\frac{dV_R}{dt} = g\beta(\varepsilon_1 V_S + \varepsilon_3 V_R)R - \mu V_R \qquad (2)$$

Transition between seasons

Having described the derivation of the *within season* model, we describe the dynamics between two growing seasons. The parameter (λ) for transfer of pathogen between seasons represents the proportion of infected material left in the field from one cropping season to the next as a result of environmental factors and farming practices (e.g. burning of stubble, ploughing, etc.). Because one aim of this study was to investigate the similarities and differences between a continuous and a seasonal model, rather than to investigate agricultural practices, we decided to keep constant across seasons the initial total crop density H_0 and the proportion θ of crop sown in the region which was resistant. Consequently, the initial total crop density and carrying capacity for each cultivar are given by

$$K_R = \theta K, \quad K_S = (1-\theta)K,$$
$$R(0) = \theta H_0, \quad S(0) = (1-\theta)H_0$$

where K is the total (i.e. over both cultivars) carrying capacity and $t = 0$ is the start of each growing season.

Furthermore, the initial density of pathogen present at the start of a season is a fraction λ of the total density remaining at the end of the previous season. $V(0)$ and $A(0)$ are initial conditions to be chosen the first season they are in the system. Then, for each subsequent growing season:

$$V(0) = \lambda V(\text{end of previous season})$$
$$A(0) = \lambda A(\text{end of previous season})$$

Numerical simulations and measures of durability of resistance

All simulations are started with only the susceptible cultivar and the avirulent strain of the pathogen. This system was solved numerically until a steady state was reached. Stabilization of the system was assessed by comparing both final densities (A and S) at the end of two consecutive seasons. The system was considered to be stabilized if both densities (A and S) differed by less than 1% when measured at the end of two consecutive seasons, that is $\frac{A(t_{i+1})-A(t_i)}{A(t_i)} < 0.01$ and $\frac{S(t_{i+1})-S(t_i)}{S(t_i)} < 0.01$ where $A(t_i)$ and $S(t_i)$ are densities of the avirulent pathogen and susceptible crop at the end of season i and the pathogen was considered as extinct for a density smaller than 10^{-14} at the end of any season.

We then assumed that the virulent strain of the pathogen was already present in very small quantities in the background of the system and could start to spread when the resistant crop was introduced. As a result, we introduced the resistant cultivar and virulent strain of the pathogen at the start of the season following stabilization of the initial system. The virulent strain of the pathogen was introduced at a very low density $V(0) = A(0)/1000$, where $A(0)$ and $V(0)$ are the densities of the avirulent and virulent strains of the pathogen, respectively, at the time of introduction of the virulent strain of the pathogen.

Two measures of durability of resistance were studied. Firstly, the widely recognized time until take-over of the virulent pathogen is defined as the time for the virulent pathogen to reach a given threshold (95% in this work) in the pathogen population, that is the time until $V/(A+V) \geq 0.95$ (van den Bosch and Gilligan, 2003). Secondly, because it is often thought that yield is related to green canopy area duration (Gaunt, 1995), we also introduced a measure of durability of resistance based on the crop estimated green canopy area duration. We calculated this as the sum, over the entire growing season, of the healthy tissue for each crop separately. Soon after the resistant cultivar was introduced, it had an advantage over the susceptible cultivar but this advantage decreased with time as the virulent strain of the pathogen developed, so that differences in green canopy area duration between the resistant and susceptible cultivars became small. As a result, we define the usefulness time of the resistant crop as the number of seasons before the estimated gain in green canopy area duration per plant due to the use of the resistant cultivar becomes negligible (<5%), that is the time until there is no advantage for the farmer to use a resistant cultivar rather than a susceptible cultivar.

Results

In this section, we successively demonstrate whether (1) durability of resistance responds to cropping ratio and compare this to results from the previously published continuous model (van den Bosch and Gilligan, 2003), (2) cost of virulence (through ε_1, ε_2 and ε_3) affects durability of resistance, (3) the two measures of durability of resistance differ and (4) the two models introduced differ.

(1) Increasing the proportion of resistant cultivar θ significantly reduces the time until take-over (Figures 1 and 2). This result was comparable to that obtained when using a continuous model. On the other hand, usefulness time is not affected by increasing cropping ratio θ.

(2) Figures 1 and 2 also show that, when the cost of virulence affects infection (model 1), take-over time decreases with increasing ε_1 and ε_3 (results not shown for ε_3) and increases with increasing ε_2. Small values of ε_1 and ε_3 and large values of ε_2 produce extreme cases where both strains of the pathogen coexist (infinite take-over time) or the virulent strain of the pathogen dies out. Coexistence, defined as an equilibrium state where the virulent strain of the pathogen neither dies out nor takes over, is therefore directly related to the cost of virulence. When usefulness time is the measure of durability of resistance, similar trends are observed, with the yield per plant for the resistant cultivar only remaining significantly greater than that for the susceptible

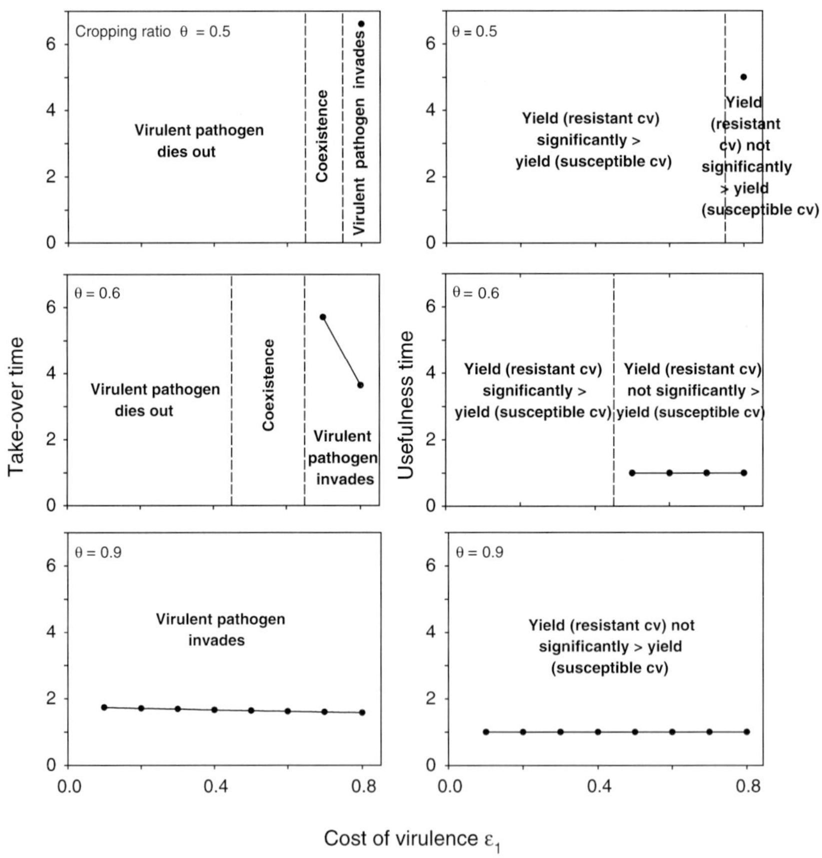

Figure 1. Effects of cost of pathogen virulence (ε_1) and cropping ratio (θ) on the two measures of durability of host resistance (take-over time and usefulness time) to when the cost of virulence is assumed to affect infection (model 1). Model parameter (Table 1) values used in these simulations are $\beta = 0.015$, $\lambda = 0.05$, $\varepsilon_2 = 0.6$, $\varepsilon_3 = 0.9$, $\mu = 30$, $g = 2000$, $r_S = 18.4$, $r_R = 18.4$, $K = 3.5$, $H_0 = 0.01$, $A(0) = 0.2$ and $V(0) = 0.001\ A(0)$. A smoothed spline was fitted to the calculated take-over and usefulness times.

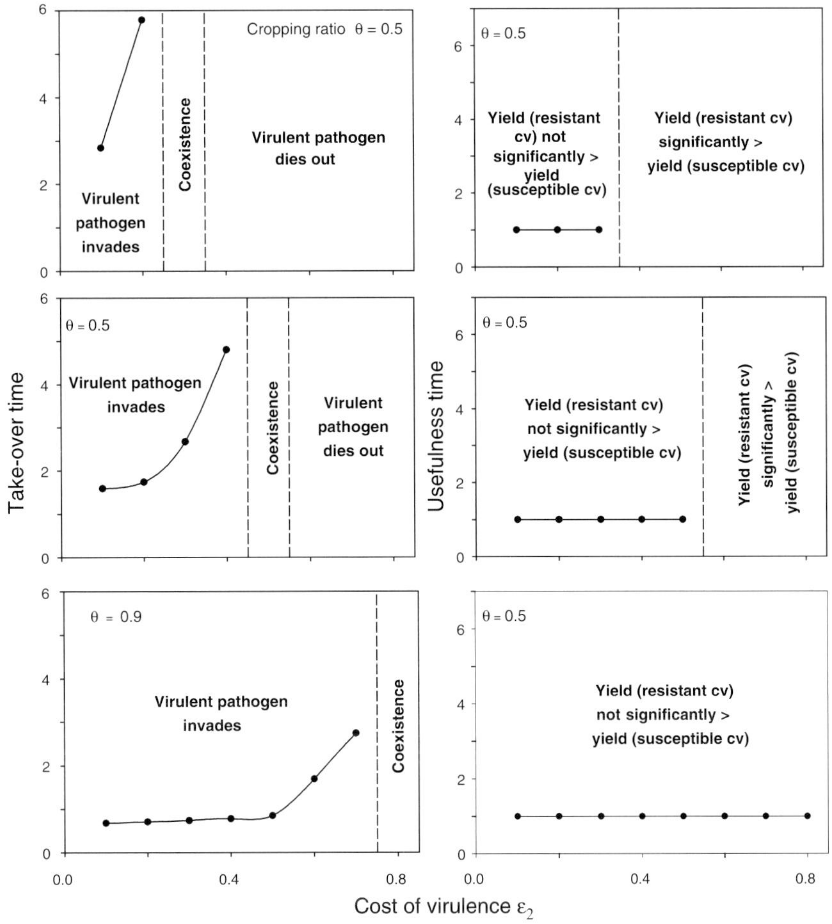

Figure 2. Effects of cost of pathogen virulence (ε_1) and cropping ratio (θ) on the two measures of durability of host resistance (take-over time and usefulness time) to when the cost of virulence is assumed to affect infection (model 1). Model parameter (Table 1) values used in these simulations are $\beta=0.015$, $\lambda=0.05$, $\varepsilon_1=0.4$, $\varepsilon_3=0.9$, $\mu=30$, $g=2000$, $r_S=18.4$, $r_R=18.4$, $K=3.5$, $H_0=0.01$, $A(0)=0.2$ and $V(0)=0.001\ A(0)$. A smoothed spline was fitted to the calculated take-over and usefulness times.

cultivar for small values of ε_1 and ε_3 (results not shown for ε_3) and large values of ε_2.

(3) The two measures of durability of resistance used in this paper are measured in the same units (number of seasons) and are therefore directly comparable. Although it might be expected that the advantage of the resistant cultivar over the susceptible cultivar would decrease as the population of the virulent strain of the pathogen increases, and that the two measures of durability of resistance are similar, Figures 1 and 2 show that this is not the case. For instance, even with the virulent and avirulent strains of the pathogen coexisting, the resistant cultivar may produce a yield (as measured by green canopy area duration) not significantly greater than that of the susceptible cultivar. Furthermore, when the virulent strain of the pathogen does take over, making the resistant cultivar 'useless', the time until take-over is always larger than the usefulness time.

(4) Although results for simulations when cost of virulence was assumed to affect the pathogen spore production (model 2) are not illustrated, there was no difference between the two models in terms of usefulness time and take-over time.

Discussion

Introduction of seasonality into the continuous model of van den Bosch and Gilligan (2003) did not qualitatively change the effect of cropping ratio on the durability of resistance. However, introduction of a cost of virulence (ε) greatly affected the dynamics of the virulent and avirulent pathogen strains and the measures of durability of resistance. The most important effect on the dynamics of the pathogen is the possibility of coexistence of the two strains for certain values of ε. Since results were qualitatively similar for the two measures of durability of resistance and similar patterns were observed for increasing ε_1 or increasing ε_3, we will concentrate on the cases where take-over time is used as the measure of durability of resistance and where cost of virulence is expressed through ε_1 and ε_2. To understand the possibility of coexistence in relation to cost of virulence, let us assume a very simple case where ε_2 is very small and ε_1 and ε_3 are equal ($\varepsilon_1 = \varepsilon_3 = \varepsilon$). If we consider an initial system in equilibrium with only the avirulent strain of the pathogen and the resistant and susceptible cultivars present, it follows that, if ε is 'large enough' and the fraction of S at equilibrium is 'not too large', the virulent strain of the pathogen will invade. Similarly, if we consider an initial system at equilibrium with only the virulent strain of the pathogen and the resistant and susceptible cultivars present, it follows that the avirulent strain of the pathogen will invade provided the fraction of S at equilibrium is 'large enough'. Thus, in a system where both strains of the pathogen are present, this shows that, for some costs of virulence (i.e. combinations of ε^s), coexistence can occur since both strains can re-invade the system if they reach small proportions of the population. This result seems in accordance with observed field data; even when a major gene for resistance has broken down, coexistence has occurred (Brun et al., 2004).

Results of this work demonstrate that the two measures of durability of resistance studied are not equivalent (e.g. both strains of the pathogen can coexist (hence no take-over) while the yield per plant of the resistant cultivar becomes significantly less than that of the susceptible cultivar). This emphasizes the importance of clearly differentiating study of pathogen populations from study of plant productivity. Although different combinations of parameters produced different values for take-over time, the usefulness time was always shorter than the take-over time. This suggests that, although the virulent strain of the pathogen may sometimes take longer to invade the pathogen population, differences in take-over time will have little effect on the changes in green canopy area duration of the plant and can be explained, similarly to the appearance of coexistence of both strains of the pathogen, as another direct consequence of the introduction of costs of virulence in the model. For instance, if we consider the very simple case where both susceptible and resistant crops are present in similar proportion, the usefulness time would be expected to be much shorter than the take-over time because the loss of usefulness of the resistant crop can be due to the infection by the avirulent strain of the pathogen. Therefore, before such results can be used in practice, further model studies are needed to confirm our findings. For instance, one factor which should be tested is the period used to calculate the green canopy area duration and usefulness time. One alternative to the solution presented in this work would be to restrict the period for calculating the green canopy area duration to later stages of growth rather than to the entire growing season. If these results are shown to be robust, one consequence is that total annual increases in yield obtained through the introduction of resistant cultivars would be directly related to the proportion of the area sown with the resistant cultivar (θ). This suggests that it might be most profitable to introduce a resistant cultivar over a large area, to produce a larger gain in *total yield*, even though this strategy would decrease the time for take-over by the virulent strain of the pathogen.

To date, many field experiments investigating durability of resistance have concentrated on pathogen population frequencies (Brun et al., 2000; Rouxel et al., 2003). This work shows that such observations may not give the best insight into the effects of a breakdown in resistance and that the host green canopy area duration should also be investigated. The results also show that cost of virulence greatly affects durability of resistance, suggesting that experiments must focus on accurately estimating such fitness cost at each stage (Li et al., 2004b; Huang et al., 2006) of the pathogen life cycle to obtain an overall cost of virulence as defined for this model.

This model can be used to investigate the effects of spatial and temporal strategies for deployment of resistance genes on durability of resistance. For instance, spatial effects can be investigated further at the local scale. Temporal effects to be analysed could include agricultural practices such as crop rotation through varying cropping ratio, possibly combined with fungicide treatments (hence varying the rate of pathogen death and cessation of sporulation μ as a function of time) to investigate the issue of inoculum thresholds (Wherrett et al., 2004). Furthermore, some of the simplifying assumptions used in this model may be relaxed. For instance, it was assumed that the cost of virulence ε was constant with time. It can be argued that such a cost may diminish as a function of time. Other areas of research to be pursued include pyramiding of major resistance genes or introducing a probability of mutation (or gene deletion) for the pathogen.

Furthermore, the model developed can be used to analyse alternative measurements of durability, in particular through economic indicators. In the economic literature (Regev et al., 1983; Hurley et al., 2001), durability is generally measured by the accumulated profit for farmers over a long period of time (say 20 years). To implement such an indicator, a damage function that describes the fact that yield loss increases with increasing density of pathogen that can survive on each crop would need to be added. The damage is small if the frequency of the virulent pathogen is very small and if the resistant cultivar is efficient in reducing the density of the avirulent pathogen strains. However, the damage becomes significant as the frequency of the virulent pathogen and the density of pathogens increase. In an economic framework, a strategy is durable if it preserves the long term economic gain without producing excessive short-term economic losses. Here, the long-term gain is generally obtained if the frequency of the virulent pathogen is kept small; the cultivar is then still efficient in controlling the damage caused by the pathogen. However, such an objective may be achieved only by preserving some avirulent pathogen strains in the short term, which would increase damage and give some short-term economic losses. Such an analysis would be worthwhile for two reasons. Firstly, most of the applied economic literature on the durability of resistance considers insect pests and, to our knowledge, no application has been made to fungal pathogens. Secondly, we expect the usefulness time indicator to be correlated with the accumulated economic profit indicator. If confirmed, this would increase the interest in using the 'usefulness time indicator'.

Acknowledgements

This work was funded by the European Union (project QLK5-CT-2002-01813, SECURE). Rothamsted Research receives Grant-Aided support from the UK Biotechnology and Biological Sciences Research Council. The French work was supported by INRA, through the grant "impact des innovations variétales".

References

Aubertot JN, West JS, Bousset-Vaslin L, Salam MU, Barbetti MJ and Diggle AJ (2006) Modelling for integrated avirulence management: Case study of phoma stem canker of oilseed rape (*Brassica napus*). European Journal of Plant Pathology 114: 91–106

Brun H, Levivier S, Somda I, Ruer D, Renard M and Chèvre AM (2000) A field method for evaluating the potential durability of new resistance sources: Application to the *Leptosphaeria maculans–Brassica napus* pathosystem. Phytopathology 90: 961–966

Brun H, Huteau V, Ermel M, Eber F, Chèvre AM and Renard M (2004) Field behaviour of oilseed rape genotypes carrying major resistance genes exposed to different *Leptosphaeria maculans* populations. International Organisation for Biological Control Bulletin 27: 95–100

Delourme R, Chèvre AM, Brun H, Rouxel T, Balesdent MH, Dias JS, Salisbury P, Renard M and Rimmer SR (2006) Major gene and polygenic resistance to *Leptosphaeria maculans* in oilseed rape (*Brassica napus*). European Journal of Plant Pathology 114: 41–52

Gaunt RE (1995) The relationship between plant-disease severity and yield. Annual Review of Phytopathology 33: 119–144

Howlett BJ (2004) Current knowledge of the interaction between *Brassica napus* and *Leptosphaeria maculans*. Canadian Journal of Plant Pathology 26: 245–252

Huang YJ, Li ZQ, Evans N, Rouxel T, Fitt BDL and Balesdent MH (2006) Fitness cost associated with loss of the *AvrLm4* avirulence function in *Leptosphaeria maculans* (phoma stem canker of oilseed rape). European Journal of Plant Pathology 114: 77–89

Hurley TM, Babcock BA and Hellmich RL (2001) Bt corn and insect resistance. An economic assessment of refuges. Journal of Agricultural and Resource Economics 26: 176–194

Kuhn ML, Gout L, Howlett BJ, Melayah D, Meyer M, Balesdent MH and Rouxel T (2006) The genome of *Leptosphaeria maculans*. European Journal of Plant Pathology 114: 17–31

Li H, Sivasithamparam K and Barbetti MJ (2003) Breakdown of a *Brassica rapa* subsp. *sylvestris* single dominant blackleg resistance gene in *B. napus* rapeseed by *Leptosphaeria maculans* field isolates in Australia. Plant Disease 87: 752

Li H, Damour L, Sivasithamparam K and Barbetti MJ (2004a) Increased virulence and physiological specialization among western Australian isolates of *Leptosphaeria maculans* breaking down existing single dominant gene-based resistance in six cultivars of *Brassica napus*. Brassica 6: 9–16

Li H, Sivasithamparam K, Barbetti MJ and Kuo J (2004b) Germination and invasion by ascospores and pycnidiospores of *Leptosphaeria maculans* on spring-type *Brassica napus* canola varieties with varying susceptibility to blackleg. Journal of General Plant Pathology 70: 261–269

Marcroft SJ, Purwantara A, Salisbury PA, Potter TD, Wratten N, Khangura R, Barbetti MJ and Howlett BJ (2002) Reaction of a range of *Brassica* species under Australian conditions to the fungus, *Leptosphaeria maculans*, the causal agent of blackleg. Australian Journal of Experimental Agriculture 42: 587–594

Marcroft SJ, Sprague SJ, Pymer SJ, Salisbury PA and Howlett BJ (2003) Factors affecting production of inoculum of the blackleg fungus (*Leptosphaeria maculans*) in south-eastern Australia. Australian Journal of Experimental Agriculture 43: 1231–1236

McDonald BA and Linde C (2002) Pathogen population genetics, evolutionary potential, and durable resistance. Annual Review of Phytopathology 40: 349–379

Pietravalle S, van den Bosch F and Evans N (2004) Durability of resistance, a modelling approach. International Organisation for Biological Control Bulletin 27: 119–123

Regev U (2002). Pest resistance in agriculture: An economic perspective. The Economics of Managing Biotechnologies (pp. 51–69). Kluwer Academic Publishers

Regev U, Shalit H and Gutierrez AP (1983) On the optimal allocation of pesticides with increasing resistance - the case of Alfalfa Weevil. Journal of Environmental Economics and Management 10: 86–100

Rouxel T, Penaud A, Pinochet X, Brun H, Gout L, Delourme R, Schmit J and Balesdent MH (2003) A 10-year survey of populations of *Leptosphaeria maculans* in France indicates a rapid adaptation towards the *Rlm1* resistance gene of oilseed rape. European Journal of Plant Pathology 109: 871–881

Turkington TK, Clayton GW, Klein-Gebbinck H and Woods DL (2000) Residue decomposition and blackleg of canola: influence of tillage practices. Canadian Journal of Plant Pathology 22: 150–154

van den Bosch F and Gilligan CA (2003) Measures of durability of resistance. Phytopathology 93: 616–625

Wherrett AD, Sivasithamparam K and Barbetti MJ (2004) Establishing the relationship of ascospore loads with blackleg (*Leptosphaeria maculans*) severity on canola (*Brassica napus*). Australian Journal of Agricultural Research 55: 849–854

European Journal of Plant Pathology (2006) 114: 117–126
DOI 10.1007/s10658-005-2501-4

Dissemination of information about management strategies and changes in farming practices for the exploitation of resistance to *Leptosphaeria maculans* (phoma stem canker) in oilseed rape cultivars

P. Gladders[1,*], N. Evans[2], S. Marcroft[3] and X. Pinochet[4]
[1]*ADAS Boxworth, Cambridge, CB3 8NN, UK;* [2]*Rothamsted Research, Harpenden, AL5 2JQ, Herts, UK;* [3]*Marcroft Grains Pathology, Grains Innovation Park, 110 Natimuk Road, VIC3400, Horsham, Australia;* [4]*Centre Technique Interprofessionel des Oléagineux Métropolitains (CETIOM), Centre de Grignon, BP No. 4, 78850, Thiverval-Grignon, France; Author for correspondence (Phone: +44-1954-267666; Fax: +44-1954-267659; E-mail: peter.gladders@adas.co.uk)*

Accepted 29 August 2005

Key words: chemical control; durable resistance, cultural control, *Leptosphaeria maculans*

Abstract

The management of phoma stem canker (blackleg disease, caused by *Leptosphaeria maculans*) is an integral component of oilseed rape production. In this paper, we discuss the information about management strategies that is disseminated in Europe and Australia. New cultivars have been introduced with improved resistance to disease, but sometimes this resistance has been overcome as new races of the pathogen have emerged. When cultivars with single major gene resistance have been introduced into areas with high inoculum concentrations, significant economic damage has been caused by new races of *L. maculans* within 2–3 years. Quantitative or polygenic resistance has also been used successfully against stem canker and offers more durable disease resistance if plant breeders and farmers deploy this resistance more effectively. Strategies to improve the durability of resistance need to be developed and tested in practice. New information on the occurrence of virulence and avirulence genes in populations of *Leptosphaeria maculans* and modelling of the durability of resistance provide opportunities for plant breeders, specialist technical organisations, cooperatives, advisory services and farmers to collaborate and better exploit cultivar resistance. Changing economic and environmental factors influence cropping practices and, if to be considered successful, management strategies must show clear financial benefits. Technology transfer will need to address all aspects of managing stem canker and other diseases of oilseed rape and using effective written, verbal and electronic methods of communication.

Abbreviations: DSS – decision support system

Introduction

Phoma stem canker (blackleg), caused by *Leptosphaeria maculans*), is one of the most economically important diseases of Brassicas world-wide. It is particularly damaging on oilseed rape (canola) in the major production areas of Europe, Australia and North America (West et al., 2001). Oilseed rape was first grown in Australia in the early 1960s (Colton and Potter, 1999) and these crops were of Canadian spring cultivars which were susceptible to phoma stem canker. By 1971, *L. maculans* had become endemic to the oilseed rape producing areas (Bokor et al., 1975; McGee and Emmett, 1977). Severe epidemics soon after crop emergence in Australia (Bokor et al., 1975) and in parts of France (Brunin and Lacoste, 1970) caused complete loss of crops. Epidemiological studies and breeding programmes were started in France during the late 1960s. The winter oilseed rape cultivar Major, registered in 1971, was the first stem canker tolerant cultivar to be widely used and had the resistance gene *Rlm4* (Delourme et al., 2006). The cultivar Jet Neuf, registered in 1978, represented an important advance, combining very good quantitative (polygenic) resistance to stem canker and specific resistance (*Rlm4*) to *L. maculans* (Delourme et al., 2006). Jet Neuf was the main cultivar used all over continental Europe during the following 10 years and has been widely used as a source of resistance in breeding new cultivars.

Phoma stem canker problems only re-emerged at the end of the 1980s when a change in European Union policy led to the introduction of double zero (low erucic acid and low glucosinolate content in seed) cultivars (Bearman, 1989). These problems were evident before the modern winter oilseed crop became established in the UK in the late 1970's and provided early awareness of the phoma stem canker disease cycle in oilseed rape in relation to components of disease management. Phoma stem canker problems soon developed in southern and eastern England, even where crops were grown for the first time (Gladders and Musa, 1980). As the areas of the UK and Australia that were cropped, respectively, to winter and spring oilseed rape cultivars expanded, government research and technology transfer programmes increased. Such research and technology transfer recorded the prevalence of pests and diseases and provided strategies for their management using cultivars and agrochemicals (Fitt et al., 1997). This led to more detailed epidemiological studies and the development of improved control strategies using resistant cultivars, fungicides and modified crop agronomy (West et al., 1999; Zhou et al., 1999; Gladders et al., 2004b). Government funding of research continues and includes partnerships with industry ('LINK funding') in the UK (Gladders et al., 2004a). In France, CETIOM has strongly supported stem canker research since the beginning of the 1990's. Government, public and European Union funding helped these efforts, notably Agence de l'environnement et de la maitrise de l'energie (ADEME), Ministère de la Recherche et de la Technlogie (MRT), Ministère de l'Agriculture et de la Pêche (MAP) Comité Technique Permanent de la Selection (CTPS), and the EU-funded projects NORDIC, IMASCORE and SECURE. In Australia, state governments and grower levies invested by the Grains Research and Development Corporation account for most research and extension funding, but several private breeding companies have also been established.

A common feature, world-wide, of damaging stem canker epidemics is the importance of airborne ascospores of *L. maculans* that are produced on crop residues (West et al., 2001). In Australia, epidemics of stem canker have been more damaging than in Europe. These epidemics prevented production of canola for approximately 15 years until suitable resistant cultivars became available (Salisbury et al., 1995). Rainfall at sowing and minimum tillage cropping practices that leave infected crop residues on the soil surface combine to produce very severe early epidemics on seedlings. Technology transfer activities have emphasised the use of resistant cultivars, isolation of new crops from fields with crop residues (Barbetti et al., 2000; Khangura and Barbetti, 2001; Marcroft et al., 2004; Wherrett et al., 2004) and use of fungicides as seed or soil treatments (Ballinger et al., 1988; Khangura and Barbetti, 2002). Little emphasis has been placed on the use of fungicide spray treatments for stem canker control because of the low input – low yield production systems in many parts of Australia (Barbetti et al., 2000).

In Europe, phoma stem canker has been more damaging in some regions of France and Germany than in the UK. Several major resistance genes have been overcome (Brun et al., 2000; Rouxel et al., 2003; Balesdent et al., 2006; Sprague et al., 2006) and improved strategies to manage stem canker are required (Aubertot et al., 2006). Technology transfer to farmers has been led by CETIOM with emphasis on understanding the development of *L. maculans* and controlling it with resistant cultivars. As in the UK, current research is directed to understanding the seasonal variation in disease development (Poisson and Pérès, 1997; Thürwächter et al., 1999) and management by improved husbandry and targeting of agrochemicals. Nevertheless, the benefits from foliar fungicide treatments are uncertain due to the long periods of dissemination of ascospores and relatively short periods of fungicide persistence. Efficient disease control has been achieved when fungicide application timings are correctly guided by spore trap records indicating prevalence of

L. maculans ascospores and forecasts of the risks of onset of phoma leaf spot development (P. Gladders, unpublished data).

Technology transfer

In Australia and Europe, the transfer of results of research and development on phoma stem canker to farmers and the wider industry was initially funded by government through programmes commissioned with state advisory services and national organisations doing cultivar testing. This was achieved through a network of plant pathologists and advisers, using local meetings with farmers, and supported by various publications and press features (Table 1). There are some differences between the UK, France and Australia in the use of written, verbal and electronic communication, but all use a range of different methods. Cultivars were promoted mainly by their breeders and agrochemicals by their manufacturers. These activities have changed over the last two decades as advice to UK and Australian farmers is now usually provided by private consultants and agrochemical distributors on a chargeable basis. In Australia, France and the UK, research results and technology transfer to farmers and industry are conveyed by levy-funded bodies (GRDC, CETIOM and Home-Grown Cereals Authority (HGCA) respectively) that receive funding from farmers based on their oilseed rape production (Table 2). Activities of these bodies include conferences and meetings, field demonstrations and publications that are available to industry, consultants and farmers. In the UK, where agrochemical usage is important, manufacturers have made a significant contribution to technology transfer. Most research projects have a component of technology transfer, although this is necessarily short-term and generally only done during the course of the project. In Australia, the Canola Association of Australia, and Oilseeds Western Australia provide links between researchers, growers, marketers, oilseed crushers and exporters in eastern and Western Australia, respectively.

Technology available for the transfer of information on disease management, continues to evolve. Advances in personal computers and the electronic media, such as the Internet, have been exploited since the late 1990s. A first step is to have technical paper documents also available on Internet (see for example www.cetiom.fr), including recommendations on how to select cultivars and agronomic practices to decrease the impact of stem canker. For other diseases, such as light leaf spot (*Pyrenopeziza brassicae*) in winter oilseed rape (Welham et al., 2004), a regional forecast exemplifies free Internet-based technology transfer (Evans et al., 2002). Individual farmers and con-

Table 1. Summary of methods used for technology transfer of information and advice on control of stem canker and other diseases of oilseed rape

Method	Format	Australia	France	UK
Written	Technical leaflets	**	**	*
	Books	*	*	*
	Specialist magazines	**	***	***
	Scientific papers	**	*	**
	Conference papers	**	**	**
	Project reports	*	*	*
	Farming press (weekly)	***	***	***
	Non-specialist Press e.g. newspapers	*	–	*
	Weekly crop reports	**	*	**
Verbal	Telephone	*	*	*
	Farmer meetings	***	**	**
	Specialist training events	*	**	**
	Scientific conferences	**	*	**
	Advisory visits	**	**	*
	On farm consultancy	***	**	***
Electronic	Email	*	*	**
	Fax	**	**	*
	Websites	**	**	**
	DSS	**	*	**
	Databases	*	*	*

The relative importance of technology transfer activities is indicated by the number of asterisks (*** is the most important).

Table 2. Funding of technology transfer activities for improved disease management of oilseed rape crops

Funding sources	Australia	France	UK
EU	–	**	*
Government (State – AUS) includes Research Councils	*	*	**
Government/industry LINK	*	*	**
Levy	***	***	***
Agrochemical industry	*	*	**
Plant breeders	*	*	*
Individual farmers/consultants	*	*	*

The relative importance of funding for technology transfer activities is indicated by the number of asterisks (*** is the most important).

sultants can access the forecast to determine the risk of damaging attacks of light leaf spot and use the forecast interactively to judge the benefits of using resistant cultivars and/or fungicides (see www3.res.bbsrc.ac.uk/leafspot). The regional forecast provides strategic guidance about the changes in disease risk from year to year. Similar forecasts are being developed for phoma stem canker, but have not yet been validated (Gladders et al., 2004a). Crop-based information is still required to guide decisions on individual fields. There has been increased emphasis on rapid transfer of information on disease development in crops to provide improved guidance on seasonal risk and the timing of fungicide sprays in autumn. In the UK, information on the development of *L. maculans* in crops in autumn has been made available by Syngenta Crop Protection UK Ltd (www.syngenta-crop.co.uk/SPAWS/). Both spore trap records and crop assessment records are updated weekly in autumn to provide improved guidance on the year-to-year and crop-to-crop variation in disease development. Disease surveys and live monitoring data are available for the England through the Defra-funded CropMonitor project (www.cropmonitor.co.uk). Recently CETIOM has released on-line "@oléovar", which is an interactive tool available to select cultivars or combinations of cultivars (www.cetiom;fr/oléov@r).

The complex interactions of pathogens, crop and environment present many challenges for technology transfer. Modelling of some of these interactions and the development of decision support systems (DSS), such as the PASSWORD project (Gladders et al., 2004a), are in progress for *L. maculans* and other pathogens of oilseed rape. DSS have the advantage of allowing wide access to guidance on disease management and for decisions to be made for individual crops in response to changing conditions. To be successful, DSS must be up to date, easy to use, reliable and cost-effective (Henriksen et al., 2000).

The impact and success of technology transfer activities may be judged in part by changes at the farm level detected in surveys. New oilseed rape cultivars are taken up by the industry in 1–2 years and most have a relatively short period of commercial production before higher yielding cultivars replace them (Hardwick et al., 2002). Similarly new fungicides have been strongly promoted and adopted rapidly in the UK (Turner et al., 2000; Hardwick et al., 2002). Technology transfer to encourage use of autumn sprays to control stem canker (and light leaf spot) and to improve fungicide spray timing has taken place since autumn 1996 (Gladders et al., 1998). At that time, few farmers were applying fungicide sprays at optimum times and they therefore derived little disease control or financial benefit from fungicide application (Gladders et al., 1998). Crop surveys in England indicated a large increase in the use of autumn sprays from 2.5% of crops treated in 1992 to 75% of crops treated in 1999 (Turner et al., 2000). Changes in timing of fungicide sprays have occurred more rapidly in oilseed rape than in wheat, probably because there are more sources of information available for fungicide treatment of wheat. This is probably due to the fact that while there are more sources of information available for fungicide treatment of wheat, sources tend to give conflicting advice (Hardwick et al., 2002). A consistent message (supported by industry partners) on the benefits of autumn fungicide sprays in oilseed rape, with clear indications of disease-induced yield loss and economic benefits (Fitt et al., 1997; Turner et al., 2002), has been important for the success of this technology transfer.

In Germany and some other European countries where winter damage is important, fungicides (often triazoles) are applied in autumn to improve winter hardiness rather than for disease control. In France, especially in the Central region, the market share of fungicide applications in autumn also increased greatly (from nearly 0% in 2002 to 9% at the national level in 2003, and from 0 to 23% in the Central region near Orléans in 2003 (CETIOM postal surveys)) in response to commercial pressure from agrochemical suppliers and in opposition to independent advice from CETIOM. This increase in use of fungicides appears to be linked to an increase in no tillage techniques. In Australia, changes in cultivars and fungicide use demonstrate the adoption of new technology by farmers. In early 2004, the Canola Association of Australia, and Oilseeds Western Australia together reported that it was no longer feasible to grow cultivars containing the major gene for resistance to *L. maculans* derived from *Brassica sylvestris* and the area sown with these cultivars decreased from 60% in 2003 to <5% in 2004. Furthermore, as a response to concerns about phoma stem canker, the uptake by farmers of fungicide treatment of seed in Australia has risen from 0% in 2000 to about 80% of all seed in 2004.

Management strategies for enhancing durability of resistance in oilseed rape

To decrease the risk that new races of *L. maculans* will overcome host resistance genes, a holistic crop management approach must be adopted by the oilseed rape industry. In many situations, a combination of cultural and chemical control measures will be required to supplement the sowing of resistant cultivars to prevent yield loss from stem canker. Oilseed rape production is expected to expand in Europe in response to economic and environmental factors affecting production of arable crops. Farmers are increasingly favouring minimum tillage practices to both improve organic matter content and soil erosion and to reduce the costs of crop establishment. Minimum tillage may also solve problems associated with increasing size of farms and reductions in farm workers. However, if this minimum tillage extends to wheat production after oilseed rape, debris-borne diseases such as phoma stem canker are likely to have a strong impact on yield and oilseed rape production may not be sustainable. Higher densities of oilseed rape cropping will potentially increase air-borne inoculum of *L. maculans* and consequently result in more severe stem canker epidemics and increased yield loss (Aubertot et al., 2006). The future of the crop in both Europe and Australia will be determined by crop yields and economic performance. The benefits of improved disease management already have a strong financial basis (Fitt et al., 1997; Barbetti et al., 2000) and benchmarks for disease, yield and economic performance could be introduced to define effective crop management strategies.

Gene deployment and management strategies to reduce the likelihood of resistance breakdown

To date, strategies for breeding stem canker resistance in oilseed rape have focused on increasing the level of stem canker resistance, with less emphasis on understanding the durability of resistance. Rapid breakdown of major gene resistance has been associated with areas of high *L. maculans* inoculum concentrations (Sprague et al., 2006). General strategies for improving the durability of resistance genes include rotation of different resistance genes in space and time, pyramiding different resistance genes and the use of multilines or mixtures of cultivars with different resistance genes (McDonald and Linde, 2002). Modelling provides some insights into the effects of different strategies (Pietravalle et al., 2006). *L. maculans* can overcome novel single major gene resistance within 3 years (Brun et al., 2000; Li et al., 2003b; Rouxel et al., 2003; Sprague et al., 2006) and has caused erosion of resistance in quantitatively resistant cultivars (Salisbury et al., 1995; Delourme et al., 2006; Li et al., 2003a, 2004). Such erosion of resistance has been reported in several Australian cultivars, such as Dunkeld, Monty, Karoo, Charlton and Ripper that are no longer grown, primarily due to their greater susceptibility to *L. maculans* than when they were first released.

In European populations of *L. maculans*, there is a high frequency of virulence alleles that can overcome resistance genes *Rlm1*, *Rlm2*, *Rlm3*, *Rlm4*, *Rlm5* and *Rlm9*, often in complex races (Balesdent et al., 2006; Stachowiak et al., 2006). This largely reflects the history of use of these resistance genes in oilseed rape and the predominance of small number of cultivars in commercial production at any one time. The cultivar Jet Neuf (*Rlm4*) was widely grown in the early 1980s, and more recently Capitol (*Rlm1*), Bristol (*Rlm2*, *Rlm9*), Express (*Rlm2*), Mendel (*Rlm3*), Falcon (*Rlm4*), Synergy (*Rlm4*) and Apex (*Rlm9*) have been widely grown in some countries. Two resistance genes, *Rlm6* and *Rlm7*, appear to still be useful for phoma stem canker control in Europe. However, there is evidence from France that *Rlm6* may be overcome in 2–3 years (Brun et al., 2001) and isolates with virulence against *Rlm7* have been detected (Stachowiak et al., 2006). Very careful management of these specific resistance genes will therefore be required if they are introduced into commercial cultivars.

Quantitative resistance or field resistance (Delourme et al., 2006) is considered to be stable in France, though it is less effective than specific (major gene) resistance. As it is a partial resistance, its effectiveness for stem canker control is sensitive to inoculum concentration. Pyramiding of resistance genes has been used successfully for control of wheat rust (Pederson and Leath, 1988) and combinations of quantitative and major gene resistance should be considered for control of phoma stem canker (Salisbury et al., 1995; Pinochet et al., 2004). It may also be appropriate to release cultivars that have more than one major resistance gene (such as Bristol) and to use cultivars with differing combinations of resistance genes. Thus cultivar diversification schemes, as used for management of various cereal pathogens, can be introduced to improve phoma stem canker management as has been proposed in France (Anon. 2004, Pinochet et al., 2004; Table 3). Cultivars of oilseed rape may be assigned to groups based on their major specific resistance genes and quantitative resistance. Within these groups (currently designated 1, 2, 3 and 4), sub-groups with the same major resistance gene or genes (e.g. cultivars with quantitative resistance and major resistance genes *Rlm1, Rlm4* or *Rlm1* and *Rlm4* could be sub-groups in Group 2) may be identified (Table 3). To exploit major gene resistance, the strategy involves careful planning of cropping so that cultivars with the same major resistance genes are not grown in successive years or in nearby fields. Where cultivars lack quantitative resistance and their major resistance genes have been overcome by *L. maculans*, they should be grown only in areas where the risk of stem canker is low. Alternation and rotation of cultivars presents some difficulties for management between neighbouring farms and may only be appropriate for use on a regional scale. With changing preferences from year-to-year, seed production would also require careful management. These strategies are untested and improved control of volunteer oilseed rape plants may be required to decrease selection pressure on the pathogen.

There may be opportunities to have localised use of resistance genes and use of cultivar mixtures or multilines (Mundt, 2002). The latter may be of greater value in Australia where secondary spread of *L. maculans* is more important than in Europe (West et al., 2001) but suitable commercial material is not yet available. Regionally deployed resistance is another option, but would be difficult to manage. To be successful, information on the nature of a new cultivar's genetic resistance should be available before its commercial release. Careful monitoring of new *L. maculans* races would be required annually after the introduction of new

Table 3. Diversification scheme to exploit different types of phoma stem canker resistance in oilseed rape cultivars

	Group 1	Group 2 (with sub-groups based on different major resistance genes)	Group 3 (with sub-groups based on different major resistance genes)	Group 4
Efficacy of resistance based on specific genes	None	Partial	Partial	High
Presence of quantitative resistance (polygenic)	Yes	Yes	None	Possible, but not identified
Field resistance rating to stem canker	Highly resistant or resistant	Highly resistant or resistant	Susceptible	Highly resistant
Durability of resistance	Stable	Some risk that specific resistance genes will be overcome	Low or already failed	Uncertain
Compatibility with other groups or sub-groups	No restriction	Compatible with Groups 1 and 4, but alternate with cultivars with different specific resistance genes from Groups 2 and 3. Ensure nearby crops do not have the same resistance genes	Compatible with Groups 1 and 4, but alternate with cultivars with different specific resistance genes from Groups 2 and 3. Do not grow in high risk areas.	Compatible with Groups 1–3, but Group 4 cultivars should only be grown in alternate years
Current cultivars* (selected examples only)	Aviso (*Rlm9*) Campala (*Rlm9*) ES Astrid (*Rlm9*) Récital (*Rlm9*)	Banjo (*Rlm1, Rlm4*) Columbus (*Rlm1*) Pollen (*Rlm4*)	Maestro (*Rlm4*) Nelson (*Rlm4*)	Caiman (*Rlm7*) Roxet (*Rlm7*)

*Data from CETIOM (Anon., 2004; Pinochet et al., 2004).

cultivars so that timely action can be taken on future production strategies.

Cultural control practices

To successfully produce oilseed rape commercially and reduce the likelihood of resistance breakdown, producers are now advised to grow their crops in ways that avoid high inoculum concentrations. Experience in France (where five genes for resistance to *L. maculans* (*Rlm1, 2, 3, 4* and *9*) have been overcome) has shown that resistance genes have remained effective in regions where inoculum concentrations are low (Rouxel et al., 2003). Therefore, growing oilseed rape crops under consistently lower inoculum concentrations is a crucial management recommendation in Australia and Europe.

Cultural control practices principally aim to decrease the number of ascospores landing on oilseed rape plant tissue. This can be achieved by burying or destroying oilseed rape stubble (allowing stubble to break down naturally within a few months) or isolating crops from stubble. However, there are some environments, such as Western Australia, where stubble residues continue to produce *L. maculans* ascospores for more than 18 months because they decay slowly in the dry Mediterranean climate. Most farmers leave 2 or 3 years between oilseed rape crops in the same field in Australia, but a 1-year break is now common in some regions. Previous studies on cultivars with quantitative resistance have shown that an isolation distance of 500 m between the current crop and the stubble of the previous year's crop is sufficient to avoid the highest concentrations of *L. maculans* inoculum in Australia (Marcroft et al., 2004) and Europe (Gossende et al., 2003). If isolation distances required are relatively small (500–1000 m), individual farmers or groups of farmers could successfully manage resistance rotation on their own properties. Rotation of oilseed rape in space and time is a practical option and farmers have adopted recommendations to sow oilseed rape crops well away from *L. maculans* inoculum.

In Australia, farmers destroy primarily by burning and in Europe, where burning is not permitted, by cultivations (Schneider et al., 2003). Chopping, slashing and/or harrowing is also used to break up the stubble. This process reputedly enhances the natural decomposition of the stubble, reducing survival of the pathogen (Bokor et al., 1975; Gladders and Musa, 1979; Turkington et al., 2000). The main limitation of actively destroying oilseed rape stubble is that most methods still leave large quantities of stubble on the soil surface. In addition, they may also have detrimental environmental side-effects if they increase soil erosion or water loss. To ensure that major sources of inoculum are managed effectively, it is essential to show significant economic advantages accruing from such practices. The costs of cultivation for burial of stubbles are considerable, but should be balanced against the value of the yield benefits from decreased disease severity. Early establishment of oilseed rape is also important so that seedlings have reached the 6-leaf stage before the main phoma leaf spot epidemic develops. Phoma leaf spotting can result in death of seedlings and early infection results in the most damaging attacks of phoma stem canker (Hammond et al., 1985; Hammond and Lewis, 1986).

Chemical control practices

In Australia, triazole fungicides, including flutriafol and fluquinconazole, are registered for the control of phoma stem canker. Flutriafol is used as a fertiliser-amended fungicide as it causes toxicity if applied directly to the seed coat. Ballinger et al. (1988) found that flutriafol increased plant height, decreased stem canker severity, increased plant survival, decreased leaf lesion severity and increased yields. Fluquinconazole is the predominant fungicide used by farmers for stem canker control due to the ease of application (seed dressing) and low cost. Seed treatment may be undervalued as it is a strategically important control measure with potential to prevent the introduction of new races of seed-borne pathogens.

There has been some interest in chemically treating crop residues to delay or inhibit ascospore production, but this has not yet been developed for use on farms (Humpherson-Jones and Burchill, 1982; Wherrett et al., 2003, 2004). In the UK, foliar fungicides have given good control of stem canker and resulted in significant increases in yield (Gladders et al., 1998, 2004a; West et al., 1999; Zhou et al., 1999). However, fungicide use is not always warranted as its economic viability will depend on yield benefits in relation to disease severity (dependent on seasonal conditions) and host resistance. Early guidance based on weather factors could be used to identify situations where chemical treatment is required (Gladders and Symonds 1995; Salam et al., 2003). In future, the role of fungicides may change if fungicide-resistant populations of *L. maculans* and other pathogens should arise or if existing product registrations are not supported in future.

Targets for technology transfer

Technology transfer for improving the management of phoma stem canker will focus on several target groups (Table 4). Oilseed rape breeders must be conversant with issues relating to the durability of resistance genes and combinations of quantitative and major gene resistance. Policy makers should be made aware of the importance of quantitative resistance for sustainable production and the need to characterise it (Delourme et al., 2006). Exploitation of resistance genes requires detailed understanding of pathogen populations and virulence genes and how these are changing in time and space. This will require further annual monitoring using methods already developed (Balesdent et al., 2006) and funding should be co-ordinated internationally. Single major resistance genes are unlikely to survive unless supported by other effective resistance genes. Strategies to ensure resistance remains durable must be economically viable if they are to influence where new cultivars are grown. Equally, farmers and their advisers need to be informed about the necessity to conserve genetic and chemical resources and the benefits of managing disease risk. Levy-funded bodies are expected to have a major role to play in supporting this technology transfer. An integrated approach to control of phoma stem canker will be required, as use of resistant cultivars must be supported by contributions from crop agronomy (including rotations and spatial distribution of crops), general hygiene to bury or reduce the impact of crop residues and fungicides. It is important to consider the economic benefits from improved control of other diseases affecting oilseed rape (e.g. sclerotinia stem rot, light leaf spot) and adapt the strategies to maximise overall benefits. While plant breeders can be targeted as a group through technical seminars and direct contact, advice to farmers and their advisers will require a more sustained approach using a variety of methods for technology transfer.

Table 4. Targets for technology transfer activities on improving disease management in oilseed rape

Target	Australia	France	UK
EU/policy	*	*	*
Government (National State – AUS)	**	*	*
Government/industry LINK	*	*	*
Levy bodies	***	***	**
Agrochemical industry	**	**	**
Plant breeders	***	***	***
Other scientists	*	*	*
Individual farmers/consultants	***	***	***
General public	*	*	*

Priorities for technology transfer activities are indicated by the number of asterisks (*** is the most important).

Acknowledgements

Funding from the European Union through the SECURE project (QLK5-CT-2002-0813) and from the Department for Environment, Food and Rural Affairs and the Biotechnology and Biological Sciences Research Council in the UK and the Grains Research and Development Corporation in Australia is gratefully acknowledged.

References

Anonymous (2004) Colza d'hiver. 40 pp. CETIOM, Thiverval-Grignon, France

Aubertot JN, West JS, Bousset-Vaslin L, Salam MU, Barbetti MJ and Diggle AJ (2006) Modelling for Integrated Management: case study of phoma stem canker of oilseed rape (*Brassica napus*). European Journal of Plant Pathology 114: 91–106

Balesdent MH, Louvard K, Pinochet X and Rouxel T (2006) A large-scale survey of races of *Leptosphaeria maculans* occurring on oilseed rape in France. European Journal of Plant Pathology 114: 53–65

Ballinger DJ, Salisbury PA, Kollmorgen JF, Potter TD and Coventry DR (1988) Evaluation of rates of flutriafol for control of blackleg of rapeseed. Australian Journal of Experimental Agriculture 28: 517–519

Barbetti MJ, Khangura R, Salam MU, Carmody P, Sweetingham M, Walton G and Zaicou-Kunesch C (2000) Managing Blackleg. Agriculture Western Australia, Bulletin 4480: 4

Bearman M (1989) The oilseed rape market in the United Kingdom. Aspects of Applied Biology, Production and Protection of Oilseed Rape and other Brassica Crops 23: 1–8

Bokor A, Barbetti MJ, Brown AGP, MacNish GC and Wood PMcR (1975) Blackleg of rapeseed. Journal of Agriculture of Western Australia 16: 7–10

Brun H, Levivier S, Somda I, Ruer D, Renard M and Chèvre AM (2000) A field method for evaluating the potential durability of new resistance sources: application to the *Leptosphaeria maculans-Brassica napus* pathosystem. Phytopathology 90: 961–966

Brun H, Ruer D, Levivier S, Somda I, Chèvre AM and Renard M (2001) Presence in *Leptosphaeria maculans* populations of isolates virulent on resistance introgressed into *Brassica napus* from *B. nigra* B genome. Plant Pathology 50: 69–74

Brunin B and Lacoste L (1970) Recherches sur la maladie du colza due à *Leptosphaeria maculans* (Desm.) Ces.Et de Not. II. Pouvoir pathogène des ascospores. Annales de Phytopathologie 3: 477–488

Colton B and Potter T (1999). History. In: Salisbury PA, Potter TD, McDonald G and Green AG (eds) Canola in Australia: the First Thirty Years (pp 1–4) Organising Committee of the 10th International Rapeseed Congress, Canberra

Delourme R, Chèvre AM, Brun H, Rouxel T, Balesdent MH, Dias JS, Rimmer SR and Salisbury P (2006) Major gene and polygenic resistance to *Leptosphaeria maculans*. European Journal of Plant Pathology 114: 41–52

Evans N, Steed JM, Welham SJ, Antoniw JF, Turner JA, Gladders P and Fitt BDL (2002) Interactive forecasting on the internet of light leaf spot (*Pyrenopeziza brassicae*) risk for winter oilseed rape. International Organisation for Biological Control Bulletin 25: 103–107

Fitt BDL, Brun H, Barbetti MJ and Rimmer SR (2006) World-wide importance of phoma stem canker (*Leptosphaeria maculans* and *L. biglobosa*) on oilseed rape (*Brassica napus*). European Journal of Plant Pathology 114: 3–15

Fitt BDL, Gladders P, Turner JA, Sutherland KG, Welham SJ and Davies JML (1997) Prospects for developing a forecasting scheme to optimize use of fungicides for disease control on winter oilseed rape in the UK. Aspects of Applied Biology 48: 135–142

Gladders P, Evans N, Fitt B, Walters K, Turner J, Northing P, Sutherland K, Campbell S, Selley A, Hall B, Ellerton D and Naylor D (2004a) The PASSWORD project: a decision support system for managing pests and diseases of winter oilseed rape in the UK. International Organisation for Biological Control Bulletin 27: 31–36

Gladders P, Jewell K and McDonough S (2004b) The contribution of cultivar resistance and fungicides to disease control in winter oilseed rape in England. International Organisation for Biological Control Bulletin 27: 51–56

Gladders P and Musa TM (1979) The development of *Leptosphaeria maculans* in winter oilseed rape and its implications for disease control. Proceedings 1979 British Crop Protection Conference – Pests and Diseases 1: 129–136

Gladders P and Musa TM (1980) Observations on the epidemiology of *L. maculans* stem canker in winter oilseed rape. Plant Pathology 29: 28–37

Gladders P and Symonds BS (1995) Occurrence of canker (*Leptosphaeria maculans*) in winter oilseed rape in eastern England 1977–1993. International Organisation for Biological Control Bulletin 18: 1–11

Gladders P, Symonds BV, Hardwick NV and Sansford CE (1998) Opportunities to control canker (*Leptosphaeria maculans*) in winter oilseed rape by improved spray timing. International Organisation for Biological Control Bulletin 21: 111–120

Gossende S, Penaud A, Aubertot JN, Schneider O and Pinochet X (2003) Evolution of soil surface oilseed rape residues and their ability to produce spores of *Leptosphaeria maculans*: preliminary results. Proceedings 11th International Rapeseed Congress, Copenhagen Denmark 4: 1166–1168

Hammond KE, Lewis BG and Musa TM (1985) A systemic pathway for the infection of oilseed rape plants by *Leptosphaeria maculans*. Plant Pathology 34: 557–565

Hammond KE and Lewis BG (1986) The timing and sequence of events leading to stem canker disease in populations of *Brassica napus* var. *oleifera* in the field. Plant Pathology 35: 551–564

Hardwick NV, Turner JA, Slough JE, Elcock SJ, Jones DR and Gladders P (2002) Oilseed rape and cereal diseases – how are farmers responding to their control. Proceedings of the BCPC Conference – Pests and Diseases 2002, 2: 903–910

Henriksen KE, Jørgensen LN and Nielsen GC (2000) PC-Plant Protection – a Danish tool to reduce fungicide input in cereals. The BCPC Conference – Pests & Diseases 2000 3: 835–840

Humpherson-Jones FM and Burchill RT (1982) Chemical suppression of the sexual stage of *Leptosphaeria maculans* on oil-seed rape and turnip seed crop straw. Annals of Applied Biology 100: 281–288

Khangura RK and Barbetti MJ (2001) Prevalence of blackleg (*Leptosphaeria maculans*) in canola (*Brassica napus*) in Western Australia. Australian Journal of Experimental Agriculture 41: 71–80

Khangura RK and Barbetti MJ (2002) Efficacy of Impact® to manage blackleg (*Leptosphaeria maculans*) in canola. Australian Journal of Agricultural Research 53: 311–321

Li H, Barbetti MJ and Sivasithamparam K (2003a) Responses of *Brassica napus* cultivars to *Leptosphaeria maculans* field isolates from Western Australia. Brassica 5: 25–34

Li H, Damour L, Sivasithamparam K and Barbetti MJ (2004) Increased virulence and physiological specialization among Western Australian isolates of *Leptosphaeria maculans* breaking down existing single dominant gene-based resistance in six cultivars of *Brassica napus*. Brassica 6: 9–16

Li H, Sivasithamparam K and Barbetti MJ (2003b) Breakdown of a *Brassica rapa* ssp. *sylvestris* single dominant blackleg resistance gene in *B. napus* rapeseed by *Leptosphaeria maculans* field isolates in Australia. Plant Disease 87: 752

Marcroft SJ, Sprague SJ, Pymer SJ, Salisbury PA and Howlett BJ (2004) Crop isolation, not extended rotation length, reduces blackleg (*Leptosphaeria maculans*) severity of canola (*Brassica napus*) in south-eastern Australia. Australian Journal of Experimental Agriculture 44: 601–606

McDonald BA and Linde C (2002) Pathogen population genetics, evolutionary potential, and durable resistance. Annual Review of Phytopathology 40: 349–379

McGee DC and Emmett RW (1977) Blackleg (*Leptosphaeria maculans* (Desm.) Ces et de Not.) of rapeseed in Victoria: crop losses and factors which affect disease severity. Australian Journal of Agricultural Research 28: 53–62

Mundt CC (2002) Use of multiline cultivars and cultivar mixtures for disease management. Annual Review of Phytopathology 40: 381–410

Pederson WL and Leath S (1988) Pyramiding major genes for resistance to maintain residual effects. Annual Review of Phytopathology 26: 369–378

Pietravalle S, Lemarié S and van den Bosch F (2006) Durability of resistance and cost of virulence. European Journal of Plant Pathology 114: 107–116

Pinochet X, Balesdent MH, Salvi F, Mestries E and Rouxel T (2004) Identification of specific plant resistance factors to Phoma (*Leptosphaeria maculans*) among winter seed rape varieties: interest for variety testing and for the promotion of a first step of a durable management of resistances. International Organisation for Biological Control Bulletin 27: 101–108

Poisson B and Pérès A (1997) Études relative à la maturation des périthèces de *Leptosphaeria maculans* sur les pailles de colza d'hiver necroses au collet. ANPP- Cinquième Conference Internationale sur les Maladies des Plantes 1: 345–352

Rouxel T, Penaud A, Pinochet X, Brun H, Gout L, Delourme R, Schmit J and Balesdent MH (2003) A ten-year survey of populations of *Leptosphaeria maculans* in France indicates a rapid adaptation towards the *Rlm1* resistance gene of oilseed rape. European Journal of Plant Pathology 109: 871–881

Salam MU, Khangura RK, Diggle AJ and Barbetti MJ (2003) Blackleg sporacle: a model for prediction of onset of pseudothecia maturity and seasonal ascospore showers in relation to blackleg of canola. Phytopathology 93: 1073–1081

Salisbury PA, Ballinger DJ Wratten N, Plummer KM and Howlett BJ (1995) Blackleg disease on oilseed Brassica species in Australia: a review. Australian Journal of Experimental Agriculture 35: 665–672

Schneider O, Roger-Estrade J, Aubertot JN and Doré T (2003) Analysis of the effects of superficial and deep soil tillage on the vertical distribution of oilseed rape stubble: contribution to the Integrated Pest Management of phoma stem canker. Proceedings of the 11th International Rapeseed Congress. Copenhagen, Denmark 3: 854–856

Sprague SJ, Balesdent M-H, Brun H, Hayden HL, Marcroft SJ, Pinochet X, Rouxel T and Howlett BJ (2006) Major gene resistance in *Brassica napus* (oilseed rape) is overcome by changes in virulence of populations of *Leptosphaeria maculans* in France and Australia. European Journal of Plant Pathology 114: 33–40

Stachowiak A, Olechnowicz J, Jędryczka M, Rouxel T, Balesdent MH, Happstadius I, Gladders P, Latunde-Dada A, and Evans N (2006) Frequency of avirulence genes in field populations of *Leptosphaeria maculans* in Europe. European Journal of Plant Pathology 114: 67–75

Thürwächter F, Garbe V and Hoppe HH (1999) Ascospore discharge, leaf infestation and variations in pathogenicity as criteria to predict impact of *Leptosphaeria maculans* on oilseed rape. Journal of Phytopathology 147: 215–222

Turkington TK, Clayton GW, Klein-Gebbinck H and Woods DL (2000) Residue decomposition and blackleg of canola: influence of tillage practices. Canadian Journal of Plant Pathology – Revue Canadienne de Phytopathologie 22: 150–154

Turner JA, Elcock SJ, Hardwick NV and Gladders P (2000) Changes in fungicide use on winter oilseed rape in England and Wales during the 1990s. Proceedings of the BCPC Conference – Pests and Diseases 2000 2: 865–870

Turner JA, Elcock SJ, Walters KFA, Wright DM and Gladders P (2002) A review of pest and disease problems in winter oilseed rape in England and Wales. Proceedings of the BCPC Conference – Pests and Diseases 2002 2: 555–562

Welham SJ, Turner JA, Gladders P, Fitt BDL, Evans N and Baierl A (2004) Predicting light leaf spot (*Pyrenopeziza brassicae*) risk on winter oilseed rape (*Brassica napus*) in England and Wales, using survey, weather and crop information. Plant Pathology 53: 713–724

West JS, Biddulph JE, Fitt BDL and Gladders P (1999) Epidemiology of *Leptosphaeria maculans* in relation to forecasting stem canker severity on winter oilseed rape in the UK. Annals of Applied Biology 135: 535–546

West JS, Kharbanda PD, Barbetti MJ and Fitt BDL (2001) Epidemiology and management of *Leptosphaeria maculans* (phoma stem canker) on oilseed rape in Australia, Canada and Europe. Plant Pathology 50: 10–27

Wherrett AD, Sivasithamparam K and Barbetti MJ (2003) Chemical manipulation of *Leptosphaeria maculans* (blackleg disease) pseudothecial development and timing of ascospore discharge from canola (*Brassica napus*) residues. Australian Journal of Agricultural Research 54: 837–848

Wherrett AD, Sivasithamparam K and Barbetti MJ (2004) Establishing the relationship of ascospore loads with blackleg (*Leptosphaeria maculans*) severity on canola (*Brassica napus*). Australian Journal of Agricultural Research 55: 849–854

Zhou Y, Fitt BDL, Welham SJ, Gladders P, Sansford CE and West JS (1999) Effects of severity and timing of stem canker (*Leptosphaeria maculans*) symptoms on yield of winter oilseed rape (*Brassica napus*) in the UK. European Journal of Plant Pathology 105: 715–728